Dominik Lutter

Models and methods used in the analysis of microarray expression data

Dominik Lutter

Models and methods used in the analysis of microarray expression data

Towards the identification of regulatory networks using statistical and information theoretical methods on the mammalian transcriptome

Südwestdeutscher Verlag für Hochschulschriften

Impressum/Imprint (nur für Deutschland/ only for Germany)
Bibliografische Information der Deutschen Nationalbibliothek: Die Deutsche Nationalbibliothek verzeichnet diese Publikation in der Deutschen Nationalbibliografie; detaillierte bibliografische Daten sind im Internet über http://dnb.d-nb.de abrufbar.

Alle in diesem Buch genannten Marken und Produktnamen unterliegen warenzeichen-, marken- oder patentrechtlichem Schutz bzw. sind Warenzeichen oder eingetragene Warenzeichen der jeweiligen Inhaber. Die Wiedergabe von Marken, Produktnamen, Gebrauchsnamen, Handelsnamen, Warenbezeichnungen u.s.w. in diesem Werk berechtigt auch ohne besondere Kennzeichnung nicht zu der Annahme, dass solche Namen im Sinne der Warenzeichen- und Markenschutzgesetzgebung als frei zu betrachten wären und daher von jedermann benutzt werden dürften.

Verlag: Südwestdeutscher Verlag für Hochschulschriften Aktiengesellschaft & Co. KG
Dudweiler Landstr. 99, 66123 Saarbrücken, Deutschland
Telefon +49 681 37 20 271-1, Telefax +49 681 37 20 271-0
Email: info@svh-verlag.de
Zugl.: Regensburg, Universität, Diss., 2009

Herstellung in Deutschland:
Schaltungsdienst Lange o.H.G., Berlin
Books on Demand GmbH, Norderstedt
Reha GmbH, Saarbrücken
Amazon Distribution GmbH, Leipzig
ISBN: 978-3-8381-1763-8

Imprint (only for USA, GB)
Bibliographic information published by the Deutsche Nationalbibliothek: The Deutsche Nationalbibliothek lists this publication in the Deutsche Nationalbibliografie; detailed bibliographic data are available in the Internet at http://dnb.d-nb.de.

Any brand names and product names mentioned in this book are subject to trademark, brand or patent protection and are trademarks or registered trademarks of their respective holders. The use of brand names, product names, common names, trade names, product descriptions etc. even without a particular marking in this works is in no way to be construed to mean that such names may be regarded as unrestricted in respect of trademark and brand protection legislation and could thus be used by anyone.

Publisher: Südwestdeutscher Verlag für Hochschulschriften Aktiengesellschaft & Co. KG
Dudweiler Landstr. 99, 66123 Saarbrücken, Germany
Phone +49 681 37 20 271-1, Fax +49 681 37 20 271-0
Email: info@svh-verlag.de

Printed in the U.S.A.
Printed in the U.K. by (see last page)
ISBN: 978-3-8381-1763-8

Copyright © 2010 by the author and Südwestdeutscher Verlag für Hochschulschriften Aktiengesellschaft & Co. KG and licensors
All rights reserved. Saarbrücken 2010

'Just look down there' said Denny.

'That seemingly endless convoy,

trailing along the dried up valley below,

look for all the world like ants.'

'They ARE ants' said his companion Minnie,

'And so are we'.

And it was true.

They were both ants,

perched on the edge of a clod of earth

no more than six inches high.

'Oh', sighed Denny sadly,

'I forgot'.

Robert Wyatt – Comicopera

Contents

Summary   7

Abbreviations   11

1. Background   13
   1.1. Transcriptome 13
      1.1.1. RNA-Types 15
         1.1.1.1. mRNA 15
         1.1.1.2. MicroRNA 16
         1.1.1.3. Further non-coding types of RNA 18
      1.1.2. Control of Transcription 18
         1.1.2.1. Chromatin structure 20
         1.1.2.2. Transcription Factors 20
         1.1.2.3. non-codingRNA 21
      1.1.3. Post-transcriptional control 22
         1.1.3.1. RNA transport and localization control 22
         1.1.3.2. mRNA degradation or turnover 23
         1.1.3.3. MicroRNAs 24
   1.2. Measuring gene expression 25
      1.2.1. Microarray technology 26
      1.2.2. Limitations 27
   1.3. Statistical methods and analysis models 27
      1.3.1. Mapping models 29
         1.3.1.1. Pairwise comparison 29
         1.3.1.2. Hierarchical clustering 31
         1.3.1.3. Support vector machines 33
      1.3.2. Mixture models 35

	1.3.2.1.	Principal component analysis	36
1.3.2.2.	Independent component analysis	37
1.3.2.3.	Non-negative matrix factorization	39

1.4. Conclusions 40

2. Analyzing M-CSF dependent monocyte/macrophage differentiation: expression modes and meta-modes derived from an independent component analysis   43

2.1. Background 43
2.2. Results and Discussion 45
    2.2.1. Signal Transduction 48
    2.2.2. Regulatory Sequences 49
    2.2.3. Differentiation, Cell Cycle 51
    2.2.4. Survival/Apoptosis 51
    2.2.5. Otherwise Classified 53
2.3. Conclusions 53
2.4. Methods 54
    2.4.1. Dataset 54
    2.4.2. Preprocessing 55
    2.4.3. JADE-based extraction of gene expression modes 56
    2.4.4. Sub-modes and meta-modes 57
    2.4.5. Mode analysis 59

3. Analyzing time-dependent microarray data using independent component analysis derived expression modes from human Macrophages infected with *F. tularensis holartica*   61

3.1. Introduction 61
3.2. Methods 63
    3.2.1. Sample preparation and expression level calculation 63
    3.2.2. Model assumptions 64
    3.2.3. ICA model 64
    3.2.4. Stability Analysis 66
    3.2.5. Grouping genes 67
    3.2.6. Biological relevance 68
3.3. Results 70

			3.3.1.	Pathways biostatistics	70
			3.3.2.	Hierarchical clustering	70
			3.3.3.	ICA analysis	71
	3.4.			Discussion	75

4. Intronic microRNAs support their host genes by mediating synergistic and antagonistic regulatory effects   77

    4.1. Introduction 77
    4.2. Results and Discussion 80
        4.2.1. Targets of similarly expressed host genes show correlated expression patterns 80
        4.2.2. MicroRNA host gene cluster and related target genes show significant correlations of their expression patterns and functional similarities 82
        4.2.3. Functional relation between host and target genes includes synergistic as well as antagonistic effects 84
        4.2.4. Host and target gene sets display enriched functional similarity . . 85
    4.3. Conclusion 88
    4.4. Material and Methods 88
        4.4.1. Microarray data and preprocessing 88
        4.4.2. Expression profile based analysis 89
        4.4.3. Intronic miRNAs and target prediction 90
        4.4.4. Functional similarity of host genes and target gene sets 91

5. Discussion   93

A. Monocyte/macrophage differentiation meta-modes   97

B. Intronic miRNAs   105
    B.1. Intronic miRNAs and host genes 105
    B.2. MicroRNA host gene cluster 108
    B.3. Functional similarity 109

Bibliography   113

Summary

All life known consists of cells. Every cell contains DNA. DNA is just a code. A code existent of four simple letters A, T, G and C. But the sequence composed of these letters contains nearly all information needed to form a complete organism as complex as a human being out of a single fertilized egg cell. And every single cell — up to a few exceptions — of one organism contains exactly the same DNA sequence as the fertilized egg, the genetic information. This genetic information belonging to a cell or organism is called a genome. This code is executed by the genes whereas a gene may contain structural, signalling or regulatory information.

Our comprehension of the genetic machinery regulating the expression of thousands of different genes controlling cell differentiation or responding to various external signals is still highly incomplete. Furthermore, recently discovered regulatory mechanisms like those mediated by microRNAs expand our knowledge but also add an additional layer of complexity. Since all genes are primarily transcribed into RNA, the genetic activity of gene differential expression can be estimated by measuring the RNA expression. Several techniques to measure large scale gene expression on the basis of RNA have been developed. In this work, data generated with the microarray technology, one of the most commonly used methods, were analyzed towards extracting novel biological regulatory structures.

In the following several aspects on the analysis of these large gene expression data will be discussed. Since this is nowadays a common task, a lot has been written about various methods in all its particulars, but often from a more technical or statistical point of view. However, the aim of a biologist planning and carrying out a microarray experiment lies on the acquisition of novel biological findings. In fact, there is still a gap between the experimentalists and the methods developing community. The experimentalists are often not too familiar with the latest fancy method based on modern statistics as it is used in e.g. information theory whereas the developing community normally does not deal extensively with current biological questions. Therefore, the author of this work tries to give an additional view on the field of microarray analysis and the applicability of diverse methods. Hence, the focus is to discuss commonly used methods towards their usage, the underlying biological assumptions and the possible interpretations, pros and cons. Furthermore, beyond ordinary differential gene expression analyses, this work also concentrates on an unbiased search for hidden information in gene expression patterns.

In the first section of chapter 1, a general overview about the main biological principles is given. The term transcriptome and its composition of several RNA types will be introduced. Furthermore the mechanism controlling gene expression will be presented. The chapter further explains the basic principles of microarray technology and also discusses the advantages and limitations of this method. Finally, by means of two different biological models, commonly used and a few more specialized and less popular analysis methods will be presented. In doing so, less emphasis is given on a complete and detailed mathematical description, but more on a general applicability and the biological outcome of these tools.

Chapter 2 extensively discusses the usage of a blind source separation technique, independent component analysis (ICA), on a two class microarray dataset. Monocytes extracted from human donors were differentiated into macrophages using M-CSF (Macrophage Colony-Stimulating Factor). By applying ICA to the data, so called *expression modes* or *sub-modes* could be extracted. According to referring biological annotations, these sub-modes were then combined to *meta modes* and elaborately discussed. In this way, several known biological signalling pathways as well as regulatory mechanism involved in monocyte differentiation could be reconstructed. Furthermore, a novel biological finding, the remaining proliferative potential of macrophages could also be identified. The results of this investigation were already published by the author [Lutter et al., 2008].

In chapter 3 again ICA was used, but in this case applied to time-dependent microarray data, and results were compared to a very common analysis method, hierarchical clustering. Time-dependent data was derived from human monocytes infected with the intracellular pathogen *F. tularensis*. Using the clustering approach, groups of genes referring to distinct timepoints were identified, and a temporal behaviour of genetic immune response could be reconstructed. In parallel, ICA was used to decompose the data into expression modes (analogously to chapter 2). These modes were then mapped on the experimental time course. Compared to the clustering results, the ICA-based reconstructed immune response was more detailed and temporal activity of distinct genes could be resolved more precisely. These findings were also published by the author [Lutter et al., 2009].

In the following chapter 4, three different microarray datasets were used to confirm a suggested regulatory mechanism. The observation that about 50% of all microRNAs in humans and mice are intronic and therefore coupled with the expression of protein coding genes, so-called host genes, allowed for the use of established large-scale gene expression measurement techniques to approximate microRNA expression. Since a single microRNA can regulate up to dozens of other protein-coding genes, the hypothesis that this expressional linkage includes an additional functional component was investigated. Using the ordinary clustering algorithm 'hierarchical clustering' and an approach based on gene annotations, this hypothesis could be basically confirmed. The main results were already outlined in a manuscript, which is currently under review.

Finally, in the last chapter, a short summary of the previous ones is given and a conclusion is drawn. A short outlook about further developments within the field of large gene expression data analysis is given and briefly discussed.

Taken together, the main contributions of this thesis are:

- This work provides an overview of the biology of gene expression and a discussion of the major analysis methods with a focus on applications.

- Based on a two-class microarray experiment, the outcome of an independent component analysis is investigated with respect to its biological relevance [Lutter et al., 2008].

- By separating time dependent microarray data into independent components, a method is presented that reconstructs a temporal regulatory network with high biological impact [Lutter et al., 2009].

- A regulatory motif of conserved microRNA functionality is confirmed, allowing for an expansion of the interpretation of gene expression data [manuscript currently under review].

Abbreviations

BSS	blind source separation
C	consensus model
Exp5	Exportin 5
fMRI	functional magnetic resonance imaging
FP	feature profile
GEM	gene expression mode
GEP	gene expression profile
GES	gene expression signatures
GO	gene ontology
GTF	general transcription factor
IC	independent component
ICA	independent component analysis
LVS	live vaccine strain
M-CSF	mononuclear phagocyte colony-stimulating factor
MeSH	Medical Subject Headings
miRNA	microRNA
mRNA	messenger RNA
NAT	natural antisense transcript
ncRNA	non-coding RNA
NMF	non-negative matrix factorization
NO	neurite outgrowth
NPC	nuclear pore complex
PC	principal component
PCA	principle component analysis
pre-miRNA	precursor miRNA
pre-mRNA	precursor mRNA

pri-miRNA	primary miRNA transcript
PT	pictar
RG	response group
RISC	RNA-induced silencing complex
RNPs	RNA binding proteins
SAGE	serial analysis of gene expression
SCD	stem cell development
SG	somitogenesis
SVM	Support vector machine
TF	transcription factor
TFBS	transcription factor binding site
TS	target scan
TSS	transcriptional start site
UTR	untranslated region

1. Background

This work addresses the analysis of large scale gene expression data. In this chapter we will outline the main biological mechanisms controlling gene expression, introduce a widely used technique to measure gene activity and discuss several commonly used analysis methods. First of all, however, since the perspective on what a gene has, progressively being changing — and still changes — during the last century, we will define the term gene as it is used in this work. According to a recently proposed definition [Gerstein et al., 2007] a gene is a union of genomic sequences encoding a coherent set of potentially overlapping functional products, either RNA or protein. A brief overview of the principle steps in gene expression and resulting gene products is given in figure 1.1.

1.1. Transcriptome

The transcriptome is defined as the collection of all gene transcripts in a cell present at one time. This includes coding messenger RNA (mRNA) as well as different types of non-coding RNA (ncRNA), with a broad variety of functions. Thus, the transcriptome can be seen as a mirror of the genetic activity of a cell. The transcription, as the initial cause of all cellular RNA (except viral RNA etc), is a complex process regulated by several mechanisms. Compared to the genome, the variety of the mRNA molecules even increases since each gene may produce several types of mRNA by alternative splicing. Furthermore, the lifespan of nearly all RNA molecules is limited and concerning mRNAs, their degradation is controlled in a complex manner. All these processes change their activity over time and directly or indirectly affect the composition of the transcriptome, resulting in a highly dynamical and complex property of living cells.

Regarding a living cell, the variety of different RNA types mirrors the multiple functions RNA is responsible for. These functions cover transfer of information (mRNA, tRNA), structural and enzymatic formations (rRNA) as well as regulatory functions (ncRNA). In this work we will mainly concentrate on one part of the transcriptome: the mRNA which can be extensively measured using microarray technology (see section 1.2). Since mRNA is the basis of translation, the production of proteins, it is therefore an indirect indicator of effective gene expression. Here we will discuss the regulatory mechanisms controlling these diverse parts of the transcriptome in more detail.

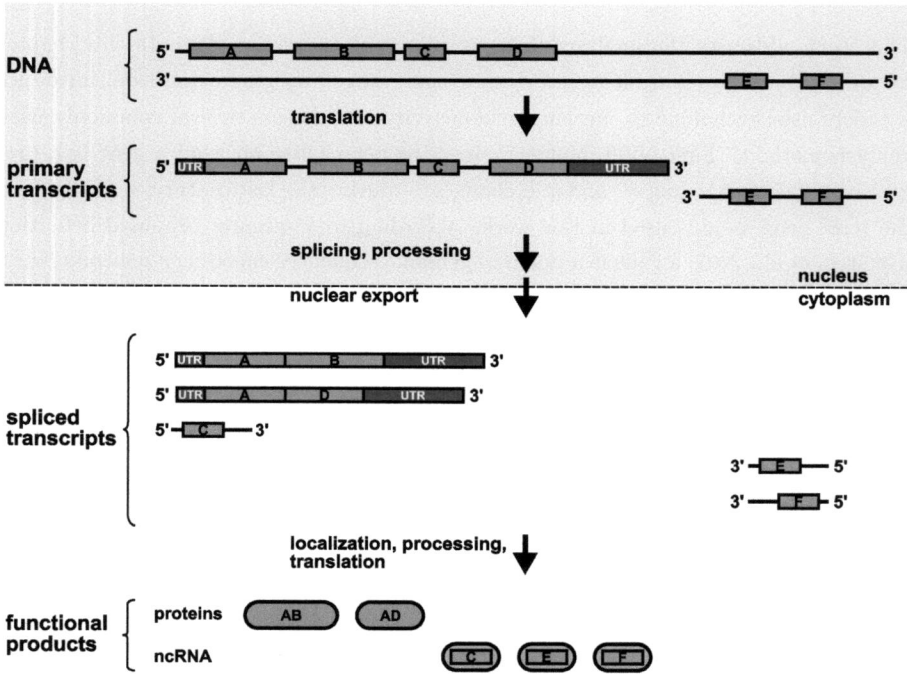

Figure 1.1.: Several steps in gene expression. From one genomic region two different primary transcripts are produced. Orange boxes (A, B and D) denote protein coding sequences, blue boxes (C, E and F) denote for non-coding gene products. After splicing and/or processing five different transcripts were produced, which finally lead to five different gene products. Proteins are indicated by rounded rectangles. Main processing steps, which can be regulated are also shown. For more details see text. The colored figure can be found at: http://www.helmholtz-muenchen.de/fileadmin/CMB/IMG/ImgDissLutter/Figure1-1.pdf

1.1.1. RNA-Types

In general RNA can be classified into two main categories: coding RNA and ncRNA. Whereas the former can be easily characterized since it contains protein coding sequences, the latter has a multitude of functional roles and is not translated into proteins. The functional roles of ncRNA are very diverse and, besides others, ncRNAs are involved in splicing, translation and gene regulation. This work is mainly based on the analysis of large scale mRNA expression profiles. In the following, we will therefore mainly discuss two RNA types, mRNA as coding RNA and microRNAs (miRNA) which have high impact on gene expression via their influence on mRNAs. Further RNA types will be briefly described in section 1.1.1.3.

1.1.1.1. mRNA

The main proportion of the mammalian transcriptome is formed by the mRNA. One mRNA always corresponds to a single gene which is defined 'as the segment of the DNA sequence corresponding to a single protein (or to a single catalytic or structural RNA molecule for those genes that produce RNA but no protein)' [Alberts et al., 2002]. The mRNA used to be primarily seen as the link between a gene and corresponding protein. This perception becomes more and more outdated since recent research supposes that mRNA functionality is more complex than expected (see chapter 4). However, in the simplest case, mRNA only transports genetic information from the DNA in the nucleus to a protein which is produced in the cytoplasm. Therefore, a gene is transcribed by RNA polymerase II into pre-mRNA and after several processing processes (see below) leaves the nucleus as mature mRNA and is then translated into a peptide by ribosomes. The mechanisms controlling transcription and translation will be discussed in sections 1.1.2 and 1.1.3. After transcription the eukaryotic precursor-mRNA (pre-mRNA) is extensively processed. Processing includes modification of the 5' and 3' end as well as 'splicing', a process to remove intron sequences from the primary transcript.

Shortly after the initiation of transcription a 5'-cap is added to the 5'-end of the mRNA by a cap-synthesizing complex associated with the RNA polymerase. The cap is exclusively added to mRNAs and helps to distinguish these from other types of RNA. Hence, it is essential for nuclear export and recognition by the ribosome. Furthermore, it prevent mRNA from degradation by RNases.

With the end of transcription an enzyme called poly-A polymerase adds approximately 200 adenosine residues to the 3'-end of the transcript. The final length of the poly-A tail is determined by so called poly-A-binding proteins, a mechanism that is so far only poorly understood. However, the poly-A tail is important for termination of transcription, export from the nucleus, the translation into protein and protection of the mRNA from degradation by exonucleases.

Protein coding sequences of eukaryotic genes are in many cases separated into small pieces, the *exons*, which are interrupted by several stretches of non-coding sequences, so-called *introns*. During RNA splicing, a process performed by the spliceosome, the introns are removed from the pre-mRNA. This is a very complex process catalyzed by a machinery consisting of five additional RNA molecules and more than 50 proteins. This modular character of a gene subdivided in several exons allows for multiple combinations of these, resulting in a variety of different mRNA molecules from one gene. Therefore, one gene is able to produce a set of different proteins, which are for instance in some case specific for different tissues [Holmberg et al., 2000]. A further interesting attribute of splicing is the generation of individual miRNAs located in intronic sequences and transcribed together with the pre-mRNA [Baskerville and Bartel, 2005]. The functions of these miRNAs will be discussed in the next sections.

1.1.1.2. MicroRNA

MicroRNAs are short, about 22nt long, noncoding RNA molecules. Since their discovery [Lee et al., 1993; Wightman et al., 1993] hundreds of miRNAs have been discovered in plants and animals [Lagos-Quintana et al., 2001; Reinhart et al., 2002; Lim et al., 2003]. After identification of their posttranscriptional gene repression by base-pairing [Hutvágner et al., 2001; Zeng and Cullen, 2003], the abundant regulatory impact on gene expression emerged. Primary expression of mammalian mRNAs is mainly subdivided into two types. One way of miRNA transcription is the transcription of miRNA genes that is controlled by an independent promoter. These genes may lead towards polycistronic miRNA transcripts with several co-expressed miRNAs [Lagos-Quintana et al., 2001; Lau et al., 2001]. The co-expression of miRNAs seems to be linked with a common function [Ambros, 2008]. The second way how a miRNA can be expressed is co-expression with protein coding genes. About half of the mammalian miRNAs, in human more than 50 %, appear to be co-expressed. These so-called *intronic* miRNAs are mainly

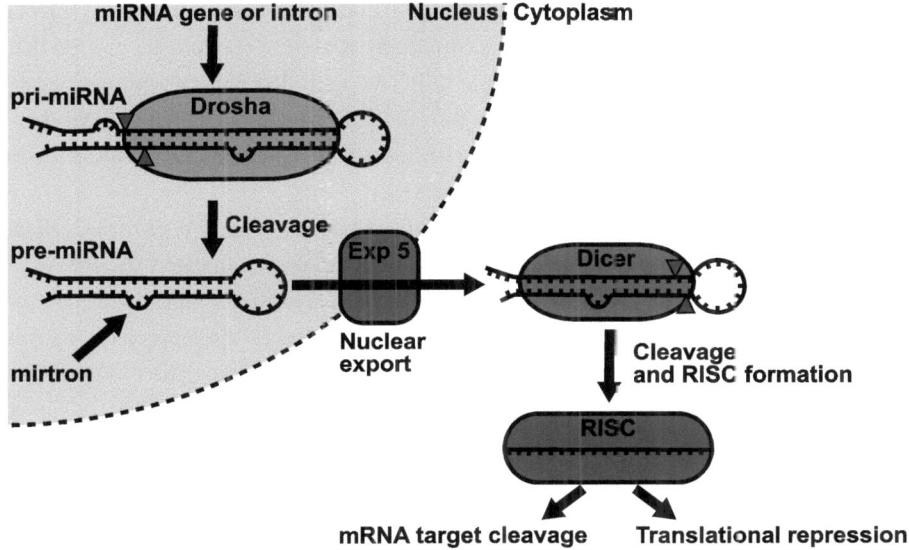

Figure 1.2.: Schematic representation of the miRNA pathway. The primary miRNA transcript (pri-miRNA) derived from a miRNA gene or an intron of a protein coding gene is cleaved by the RNase III enzyme Drosha. After this cleavage, the precursor miRNA (pre-miRNA), which may also be derived from so-called mirtrons is exported into the cytoplasm. Dicer, another RNase III enzyme cleaves the pre-miRNA and the so-called guide strand is incorporated into RISC. For a detailed explanation of the miRNA pathway see text.

located within the intron of the host genes, but miRNAs located in exons as well as in 3'UTRs (untranslated regions) have also been discovered [Lagos-Quintana et al., 2003; Rodriguez et al., 2004]. The conserved linkage of expression between a protein coding gene and a miRNA strongly suggests that there is also a functional relationship between host gene and miRNA. This could be already shown for two individual miRNAs [Barik, 2008; Zhu et al., 2009]. A general functional relationship between host genes and their intronic miRNAs is extensively analyzed in chapter 4. However, most miRNAs are therefore transcribed by RNA polymerase II, aside from some human miRNAs within alu-repetitive elements, which can be transcribed by RNA polymerase III [Borchert et al., 2006].

Maturation of miRNAs occurs through sequential processing steps. After transcription canonical primary transcripts (pri-miRNAs) forms ~70nt duplex like hairpin-loops, which are cleaved in the nucleus by the RNase III enzyme Drosha. In case of intronic miRNAs Drosha cleavage was shown to occur closely related to the splicing process [Kim and Kim, 2007]. A special type of intronic miRNAs, so called mirtrons were processed within an alternative pathway. These, also intronic miRNAs, mimic hairpin structures of pre-miRNAs and bypass Drosha-mediated cleavage to enter the miRNA pathway during splicing [Ruby et al., 2007; Berezikov et al., 2007].

After export of the miRNA precursor (pre-miRNA) from the nucleus to the cytoplasm another RNase III enzyme called Dicer mediates the next processing step [Zamore et al., 2000; Ketting et al., 2001]. Dicer recognizes the double-stranded portion of the pre-miRNA, cuts both strands of the duplex and thereby removes the loop of the hairpin. According to the current model, the end of the pre-miRNA defining the mature $\sim$22nt long miRNA is defined during nuclear cleaving by Drosha [Lee et al., 2003]. The so-called guide-strand is then selected by the Argonaut proteins and integrated into a ribonucleoprotein complex, known as the RNA-induced silencing complex (RISC). The active RISC, the complex bound to single-stranded miRNA, identifies target mRNA sequences based on complementarity and controls their expression by either degradation or inhibition of translation. A schematic representation of the miRNA pathway is shown in figure 1.2. This regulatory mechanism will be discussed in more detail in section 1.1.3.

1.1.1.3. Further non-coding types of RNA

Besides mRNAs and the already discussed noncoding miRNAs several other ncRNA types are known. They form a diverse group of RNAs including functionally well-understood RNAs such as tRNA and rRNA, as well RNA types with more or less obscure functionality, like small nuclear RNAs (snRNAs), piwi-interacting RNA (piRNA), and long non-coding RNAs (long ncRNAs). Since the impact on gene expression of most of these RNA types is marginal, unknown or not measureable by the methods discussed in this work, only a short overview of known members of the eukaryotic ncRNA family can be found in table 1.1.

1.1.2. Control of Transcription

The process of copying one DNA strand into a complementary RNA strand by the RNA polymerase enzyme is generally called transcription. In eukaryotes, several RNA poly-

RNA type		Function
miRNA (microRNA)	(mi-	(~22nt length) gene regulation by translational repressing or mRNA degradation
siRNA (small interfering RNA)		(20-25nt length) RNA interference; taming of transposons and combating viral infections
piRNA (Piwi-associated RNA)		(25-30nt length) essential in the development of germ cells
rRNA (ribosomal RNA)		mediates decoding of mRNA to amino-acid sequences of protein
tRNA (transfer RNA)		transfers a specific amino acid to a growing polypeptide during translation
snRNA (small nuclear RNA)		involved in several processes in the nucleus, like splicing and maintenance of the telomeres
long-ncRNA (longer non-coding RNA)		participate in various cellular processes, including splicing and ribosome biogenesis

Table 1.1.: Members of the ncRNA family, abbreviations and function.

merases synthesize the different types of RNA. mRNA and most miRNAs for instance are synthesized by RNA polymerase II. The whole process can be subdivided into three main stages: initiation, elongation and termination. Initiation summarizes the binding of the RNA polymerase enzyme to the DNA by recognition of the promoter, separation of the duplex DNA structure and initiation of the RNA synthesis process. During elongation, the RNA polymerase traverses the template strand from 3' to 5' generating the RNA copy of the coding strand in 5' to 3' direction. Elongation continues until the RNA polymerase encounters a termination signal encoded on the DNA. The transcription stops and the polymerase releases the DNA template as well as the newly synthesized RNA.

Transcription is the first step in gene expression and is controlled by several complex mechanisms. In the following, the main mechanisms will be briefly discussed.

1.1.2.1. Chromatin structure

In eukaryotic cells, the DNA is usually organized to a complex in the nucleus called chromatin. The material of which chromosomes are made of. It is build of DNA, histone and non-histone proteins, subdivided into nucleosomes. Transcription of a gene is strongly dependent on the structure of the chromatin. Important local alterations influencing transcription are histone modifications and nucleosome remodeling.

Histones can be modified through at least eight different ways [Kouzarides, 2007], which all have influence on transcriptional activity. For instance, histone acetylation catalyzed by histone acetyltransferases alter the chromatin structure in a way that allows for greater accessibility of the DNA. Hence DNA polymerase and transcription factors have easier access to promoter regions. In contrast, histone methylation inhibits translation through several different mechanisms [Singal and Ginder, 1999]. Since histone methylation patterns are heritable after cell division, its role during differentiation processes is very important. Furthermore histone methylation seems to have remarkable impact on the epigenetic memory [Callinan and Feinberg, 2006]. Nucleosome remodeling is mediated by chromatin remodeling complexes which also allows for greater accessibility of DNA packed in chromatin to other proteins.

In general the chromatin structure controls gene expression on a basal level. It is primarily accountable for accessibility of the DNA, thus protein coding as well as regulatory sequences. It can further influence expression rates in various ways, thereby forming the basic mechanisms of the gene expression regulatory machinery.

1.1.2.2. Transcription Factors

A protein that binds to the DNA or as a co-factor to the polymerase-DNA-complex is termed a transcription factor (TF) if it is somehow required for initiation or regulation of transcription in eukaryotes. There are general transcription factors (GTF) that are required by the RNA polymerase II for DNA binding and initiation of the RNA-synthesis. Not all of these GTFs actually bind to the DNA but are part of the huge protein complex which directly interacts with the DNA and the DNA polymerase.

Further DNA-binding proteins influence transcription in a variety of ways. They can stabilize or block the binding of DNA polymerase, directly or indirectly catalyze the acetylation or deacetylation of DNA (see 1.1.2.1) or recruit co-activator or co-repressor proteins. TFs bind DNA at either promoter sequences or cis-regulatory elements [Gill, 2001]. A promoter is defined as the nucleotide sequence in the DNA to which RNA polymerase binds and starts transcription. Promoters are found upstream of the transcriptional start site (TSS) and can include regulatory elements several kilobases away from TSS. Besides the core promoter, required to properly initiate transcription with the RNApol binding site, it mainly consists of specific TF binding sites (TFBS). Cis-regulatory elements are short DNA sequences with specific TFBSs which can be located many kilobases away from TSS. Together these sequences can be termed a 'gene control region'.

As the number of GTF is relatively small and similar for all polymerase II transcribed genes, the amount and composition of additional regulatory proteins is different for each gene. About 5-10% of all mammalian protein-coding sequences, of estimated 20,000 to 25,000 human genes [Carninci and Hayashizaki, 2007], are proposed to serve as regulators of gene transcription [Wilson et al., 2008]. Expression of each gene is controlled by a set of different TFs, whereas each of those are in turn regulated by its own set of gene regulatory proteins. The resulting exceedingly complex network controlling the expression of mammalian genes allows for a diversity of spatial and temporal different transcriptional expression patterns.

1.1.2.3. non-codingRNA

In addition to transcriptional control mechanisms based on chromatin structure or mediated by gene regulatory proteins, several ncRNAs have a functional role as regulators of transcription [Carninci, 2008]. To date, there are several molecular mechanisms identified, most of them only poorly understood. However, their positive or negative influence on the transcription rate is shown [Morris et al., 2008].

Among the various mechanisms identified by several studies are natural antisense transcripts (NAT) and the specific binding to transcription factors and or DNA sequences directly. Besides trans-NATs which mainly do not affect transcription itself (e.g. miRNAs, see 1.1.3.3), cis-NATs for instance, may inhibit transcription by histone modification within promoter regions [Osato et al., 2007]. Other ncRNAs can bind proteins involved in transcription, thus influencing their activity [Storz, 2002]. Detailed explanations and further examples can be found in a variety of recent articles [Barrandon et al., 2008; Carninci et al., 2008; Mattick and Makunin, 2006].

1.1.3. Post-transcriptional control

Gene expression starts with transcription, which produces primary RNA transcripts and is followed by several maturation steps. As shown above transcription is controlled by various different mechanisms, while in principle each step can be regulated independently. The single steps include processing of the primary transcripts, splicing and export from the nucleus to the cytosol, where their cellular localization can also be regulated. Furthermore, transcripts in the cytoplasm may be selectively destabilized, activated, inactivated or degraded. Translation, the process in which mRNA is finally translated into protein is also extensively regulated.

All these regulatory mechanism that follow transcription and affect gene expression are referred to as **post-transcriptional control**. In this chapter we will briefly discuss the main mechanisms of post-transcriptional control with strong impact on the composition of the transcriptome and gene expression. Therefore, we will basically focus on mRNA and miRNA.

1.1.3.1. RNA transport and localization control

In eukaryotic cells synthesis and diverse pre-processing steps of RNA take place in the nucleus. Several of the produced RNA types, including mRNA and pre-miRNA, are exported to the cytoplasm where they serve as a template for protein synthesis or influence the same in various ways. In general every RNA exported from the nucleus must pass through the nuclear membrane via nuclear pore complexes (NPC), but the distinct nuclear export pathways for different RNA types vary [Cullen, 2003]. As far as the exact mechanisms are understood, nuclear RNA export is highly selective and is mainly mediated by a protein family termed exportins (karyopherins). These exportins depend on the activity of a small co-factor, the GTPase Ran [Allen et al., 2000]. In case of Drosha-processed pre-miRNAs Exportin5 (Exp5) forms a heterotrimer with Ran and pre-miRNA, whereas the binding of Exp5 depends on the RNA structure but not on the sequence. After passing the NPC Ran-GTP is hydrolyzed to Ran-GDP and the pre-miRNA is released [Cullen, 2004].

In the cytoplasm pre-miRNAs undergo a final processing step: Dicer, a RNase III enzyme, binds the double stranded pre-miRNA and cuts both strands of the stem loop, generating a $\sim$22 nucleotide miRNA duplex. One strand is incorporated into RISC, whereas the other miRNA* strand is typically degraded [Bushati and Cohen, 2007].

In contrast mRNA export does not depend on Ran and karyopherins but depends on various other RNA binding proteins. Furthermore, the NPC recognizes and transports only completely processed mRNAs. Presumably, the recognition depends on cap-binding, poly-A-tail and further binding of appropriate proteins. Key proteins mediating the export of mRNA are Tap and a small co-factor termed Nxt (p15) that form a heterodimer. However, by recruitment of further proteins like UAP56 and RNA-dependent ATPases the ribonucleoprotein complexes (RNP complexes) is recognized by NPC and the intron free mRNA is exported to the cytoplasm [Iglesias and Stutz, 2008].

An exported mRNA binds to ribosomes, which translate it into a polypeptide. Some mRNAs are directed to specific intracellular locations. The direction is controlled by specific sequences mainly within 3' UTR, but also in the 5' UTR, recognized by RNA binding proteins (RNPs). These transport RNPs engage with cytoskeletal motors for directed transport. During transport several mechanisms, presumably including small non-codingRNAs and further RNA binding proteins inhibit the translation of transported mRNA [Besse and Ephrussi, 2008]. Beyond this spatial component, a temporal regulatory impact of these mechanisms controlling gene expression is assumed.

1.1.3.2. mRNA degradation or turnover

The protein production is further regulated by the mRNA lifespan. In general mRNA molecules are unstable and consistently degraded. Different eukaryotic mRNAs have different half-lives, ranging from several minutes to more than 10 hours (β-globulin mRNA) [Alberts et al., 2002]. Several independent mechanisms control mRNA turnover. Besides the common pathway, that is deadenylation followed by exosome complex mediated degradation, there is also cleavage by sequence-specific endonucleases or cleavage in response to the binding of complementary small interfering RNA (siRNAs) or miRNAs [Parker and Song, 2004].

Nearly all ~200 bp long poly-A-tails of eukaryotic mRNAs are continuously shortened by a variety of deadenylases in a 3' to 5' direction. Once the tail reaches a critical length, the 5' cap is removed and the mRNA is rapidly degraded. Decapping allows for additional digestion in 5' $\rightarrow$ 3' direction by exonucleases. Furthermore, after deadenylation the exosome, a huge protein complex containing multiple exoribonucleases [Newbury, 2006], degrades mRNA from the 3' end. This protein complex is also involved in nonsense-mediated decay [Lejeune et al., 2003; Lehner and Sanderson, 2004], a mechanism detecting nonsense mutations and prevents the production of truncated or erroneous proteins by RNA degradation. The rate of poly-A tail shortening varies from mRNA to mRNA and depends on several RNA-binding molecules which can decrease or increase the rate of deadenylation.

The cleavage of mRNA is mainly controlled by siRNA. Short double-stranded RNA molecules processed by Dicer and integrated into RISC, bind to complementary mRNA sequences and induce enzymatic cleavage [Moazed, 2009]. This process is strongly related to miRNA mediated translational control and will be discussed in detail in the next chapter.

Many untranslated mRNAs assemble in related mRNPs that accumulate in specific loci termed *P bodies* [Parker and Sheth, 2007]. P bodies interact with the decay machinery and associated mRNAs can either be degraded after decapping, remain in the P body state or reentry translation. Although many questions concerning the function of P bodies are unclear, their role in modulation of gene expression is indisputable.

1.1.3.3. MicroRNAs

Shortly after their discovery in the 1990s, the interest in miRNAs extremely increased due to the discovery of their impact on protein coding gene expression. After a miRNA is embedded into RISC, it binds to specific sequences mainly in the 3' UTRs of mRNAs and inhibits translation or causes degradation initiated by cleavage of the poly-A-tail [Grosshans and Filipowicz, 2008].

Recognition of target sites depends on extensive complementary pairing but does not require a complete match over the full miRNA length. Most miRNA binding sites identified so far include a complete 7-8mer pairing in the 'seed' region of the miRNA. This region is defined as the nucleotides 2-7 from the 5' end of the miRNA [Bartel, 2009]. Beside these canonical seed-matched sites several 6mer pairing sites and even seed mismatch sites are verified to be functional [Brennecke et al., 2005]. However, sites with insufficient 5' pairing seem to require strong 3' pairing, indicating that besides pairing the free energy also affects the stability of the miRNA:mRNA duplex [Doench and Sharp, 2004].

MicroRNAs loaded into RISC modulate gene expression mainly by downregulation of the rate of translation. This can be achieved by two different mechanisms: mRNA cleavage and translational inhibition. Cleavage of mRNA depends on sufficient complementarity of the miRNA and is identical to the siRNA pathway. In animals where miRNAs target mRNAs mainly by an imperfect match the latter mechanism, which leads to translational repression, outbalances. Two different modes of repression are currently discussed. Repression of initiation of translation and repression of elongation of the polyaminoacid chain [Cannell et al., 2008]. However, recently it has been shown that miRNAs can also activate translation of target mRNA [Vasudevan et al., 2007].

Furthermore, repression of target activity can be classified into three main categories: 'Switch', 'fine tuning' and 'neutral' [Flynt and Lai, 2008] Whereas switch refers to a inhibition of protein synthesis towards a target inactivity, tuned targets still produce functional proteins but in a lower amount. Functional miRNA:mRNA interactions without advantageous nor adverse consequences are denoted as neutral, since their effect on the phenotype is negligible. Differentiation between tuning and switch depends on the impact of translational repression. Properties modulating the impact are characteristics of the seed, GC-content (guanine-cytosine content) and the number of functional binding sites within the 3' UTR [Baek et al., 2008].

Like TFs miRNAs are affecting their target genes in different miRNA combinations and a single miRNA can target up to hundreds of different mRNAs [Betel et al., 2008]. As a consequence, the combinatorial scope allows for complex regulatory networks controlling the expression of thousands of protein coding-genes. Considering that also TFs are targets of miRNAs and in turn control their transcription, too, extensive linkage between both regulatory networks holds for multiple sources of information to control expression of individual transcripts. So far, little is known about global and local structures of these networks but recent studies provide more and more insight into the architecture and components or motifs it is composed of [Shalgi et al., 2007; Tsang et al., 2007; Yu et al., 2008].

1.2. Measuring gene expression

In the last chapter the mammalian transcriptome was briefly introduced and the most prominent RNA types were discussed. Furthermore, we discussed the main regulatory mechanisms controlling the expression of genes. In this chapter we will shortly discuss several methods that are used to measure gene expression based on RNA levels. In principle one can differentiate between methods measuring the expression of single RNA molecules or large scale methods, which are able to measure the expression of thousands of genes at once. In this work we exclusively focus on the analysis of high throughput expression data. The most commonly used method to measure large scale gene expression is the microarray technology [Kawasaki, 2006]. Further methods are serial analysis of gene expression (SAGE) [Anisimov, 2008] and Deep sequencing [Wang et al., 2009].

In the following sections the principles of microarray technology will be introduced and the applicability as well as the main issues and restrictions will be discussed.

1.2.1. Microarray technology

A microarray or genechip is a tool which allows to measure the expression of thousands of genes simultaneously. Although different techniques exist the technical principle is mainly identical. On a small support, consisting of a membrane or glass slide, probes are immobilized by covalent bonds to a chemical matrix. These probes can be short DNA fragments, cDNA or oligonucleotide sequences organized in so-called spots, complementary to nucleotide sequences of known transcripts. In spotted arrays probes are synthesized prior to deposition on the array surface and are then 'spotted' onto glass. In oligonucleotide microarrays, the probes are mostly synthesized directly onto the support.

Fluorescent-labeled cDNA molecules derived from isolated mRNA from each cell type studied are then hybridized to the genechip. Within spotted arrays one often hybridizes control and sample cDNA or cRNA labelled with two different fluorescent dyes onto one chip, whereas in oligonucleotide arrays only one color channel is used. Control and sample RNA are therefore hybridized to different chips. The measured fluorescence intensities for each spot mirrors the relative expression of the corresponding transcripts. Changes in gene expression can be estimated by computational comparison of the measured expression levels.

In this work only one channel oligonucleotide microarrays as manufactured by Affymetrix were used. Further reading about technical background, probe level data and probe annotation can be found in [Affymetrix, 2001; Irizarry et al., 2003; Liu et al., 2003]. After several normalization and preprocessing steps huge data sets of gene expression are obtained [Sarkar et al., 2009]. Typically one denotes the columns as the samples or *gene expression profiles* (GEPs) and the rows, representing the expression level of each gene across all experimental conditions. The proper analysis of such data is an elaborate task and will be extensively discussed in the next sections.

1.2.2. Limitations

Microarray technology benefits from its high throughput characteristics, but unlike methods like SAGE and Deep sequencing, it is a closed method that is limited to the genes that are represented on the chip. However, not all genes or transcripts are known yet or sequences are wrongly identified during genome annotation. A further disadvantage, compared to gene expression profiling methods like QFCR (quantitative PCR), is that it lacks accuracy. The main reason for impreciseness in measuring the expression of a particular transcript is caused by cross-hybridization, annealing of only partially complementary sequences. Furthermore, probes designed from genomic EST information may be incorrectly associated with a transcript of a specific gene.

Since a particular probe is mainly designed to match parts of the sequence of known or predicted open reading frames, different splice forms of a single genes can not be determined. Moreover, genechips only detect mRNA levels. As described above, these are subjects to comprehensive post-transcriptional regulatory mechanisms and though not obligatory translated into protein. These restrictions to gene expression stay obscure within a microarray experiment.

1.3. Statistical methods and analysis models

The first sections in this chapter contained a brief summary of the regulation and composition of the mammalian transcriptome. Several regulatory mechanisms and their interactions were described, to show the complexity of gene expression regulation. In the last section microarray technology, a widely used method that allows for the simultaneous measurement of the activity of thousands of different genes, was introduced. Microarray experiments produce high-dimensional data with little replication, thereby causing several problems of statistical analysis. The complexity and huge amounts of data pose for several bioinformatic challenges, ranging from pre-processing steps like background correction, data normalization and filtering over to gene annotation and data warehousing [Autio et al., 2009; Hackstadt and Hess, 2009; Stekel, 2003].

In this work we mainly focus on the statistical analysis of pre-processed gene expression data. The goal is to extract meaningful biological information. Typical biological goals addressed by microarray experiments include the identification of co-expressed genes, identification of genes or groups of genes with expression patterns related to experimental conditions (chemical treatment) or different cell types (tumor vs wild type), or the identification of regulatory relationships (TF - target gene).

In the field of microarray data analysis a lot of different statistical tools and methods have been developed to achieve the above mentioned goals of the biological tasks. A common classification of these methods is the distinction between supervised and unsupervised methods. Supervised methods use prior knowledge about samples or genes to extract patterns or features specific to a given class or to classify samples or genes [Lutter et al., 2006]. In contrast, unsupervised methods screen the data for interesting novel biological regularities or relationships. Additionally, one can also classify analysis methods as clustering methods, projection methods or graphical model based approaches. However, all these methods are widely discussed and precisely explained in a number of articles, reviews and books [Quackenbush, 2006; Dougherty et al., 2005; Allison et al., 2006; Berrar et al., 2003].

In the following, we will discuss several analysis methods based on the underlying biological model conceptions. Concerning the biological background one can distinguish between two main models: mapping models and mixture models. Mapping models are based on the assumption that each measured gene expression profile corresponds to a specific cellular state, chemical treatment or experimental condition, whereas mixture models are based on the assumption that a gene expression profile is composed of several biological processes running in parallel. Each process is responsible for a particular expression profile. In the following, these model conceptions and the corresponding statistical tools used in this work will be discussed. The applicability of these tools on microarray data and the biological questions that give rise to the use of a particular analysis method will be discussed below.

As mentioned above, only Affymetrix oligonucleotide gene chips were used in this work. Therefore, the following methods mainly refer to one channel gene expression profile data. However, most of these methods can be applied in a slightly modified way to two channel data as well.

1.3.1. Mapping models

Typically a microarray measurement is considered as a map of the cellular gene expression, based on mRNA levels, at a distinct time point and under certain — inner and outer — conditions. Inner conditions may refer to a developmental stage or alteration in the genotype, whereas outer conditions may be chemical treatments, starvation or physical stress. Different conditions cause the cell to react with a modification in gene expression. Changes in expression patterns can be interpreted as the phenotypic expression of regulatory mechanisms. For instance, comparing expression profiles of a TF knock-out experiment to wild type profiles will produce a list of up and downregulated genes, which can be interpreted as negatively or positively regulated TF target genes. Moreover, the differences in the temporal expression profiles of differentiating cells provide information about the activated or inactivated pathways.

Based on these model assumptions several statistical methods have been established that generate interpretable biological results. The most commonly used of these methods will be discussed in this work with regard to the above mentioned underlying biological mechanisms. In the following the expression value of a gene k in the nth of N experiments is written as x_{kn}. Two different experimental conditions can be denoted as "+" and "−", which reads then as $x_k(+)$ and $x_k(-)$ the expression of a gene k under two conditions for instance as *treatment* and *control*.

1.3.1.1. Pairwise comparison

The most canonical approach in the analysis of different gene expression patterns is to look for differentially expressed genes. The goal is to identify genes changing their expression significantly from one state to another. Dependent on the size of the dataset several methods are commonly used to identify these genes [Cui and Churchill, 2003]. Three of these will be exemplary listed and shortly specified.

- A **fold change** denotes the relative change in gene expression between two distinct experimental conditions ±. For a gene k it depends on the log-ratio

$$SignalLogRatio_k = \log_2 \frac{x_k(+)}{x_k(-)} \quad (1.1)$$

If replicates for the conditions are available one typically uses the estimated means

$x_i = \overline{x}_i(\pm)$. The fold change for gene k can then be defined as

$$FoldChange_k = \begin{cases} 2^{SignalLogRatio_k}, & SignalLogRatio_k \geq 0 \\ -2^{-SignalLogRatio_k}, & SignalLogRatio_k < 0 \end{cases}. \quad (1.2)$$

The fold change is not a statistical test, and does not provide any associated value that can indicate the level of confidence. Furthermore it is subject to bias caused by improperly normalized data or outliers.

- The **t-test** is a simple statistical test to detect differentially expressed genes. It compares two distributions, assumed to be Gaussian, to test whether the means are different. Applied to a two class microarray experiment it can be used to determine significantly differentially expressed genes. The power of the test depends on the number of samples, and therefore, is low for microarray experiments where the sample size is typically small. Furthermore, it may suffer from the same bias as the fold change if the error variance is not truly constant for all genes.

- Significance analysis of microarrays (**SAM**) is a further, widely used method to determine differentially expressed genes [Tusher et al., 2001]. It assigns a score to each gene, relative to the standard deviation of repeated measurements, based on changes in expression between two conditions. The algorithm estimates a false discovery rate (FDR) using permutations of the replicates that can be used to adjust a threshold to identify significantly regulated genes. The test is more robust for small sample sizes then the t-test, and does not assume normal distributions.

However, all of these methods only allow for a pairwise comparison of two different conditions. They rank genes accordingly to their change in expression and — if applicable — provide a significance measure. The biological meaning of these lists has to be interpreted carefully. Depending on the quality of the data or normalization errors false positives may occur. Furthermore, one can not distinguish between direct or indirect regulatory effects and, since cells react in many different ways on different treatments, genes showing high differential expression do not necessarily share a common function. Finally, these methods imply a relationship between differentially expressed genes and the experimental treatment. But the strength in alteration of expression does not depend

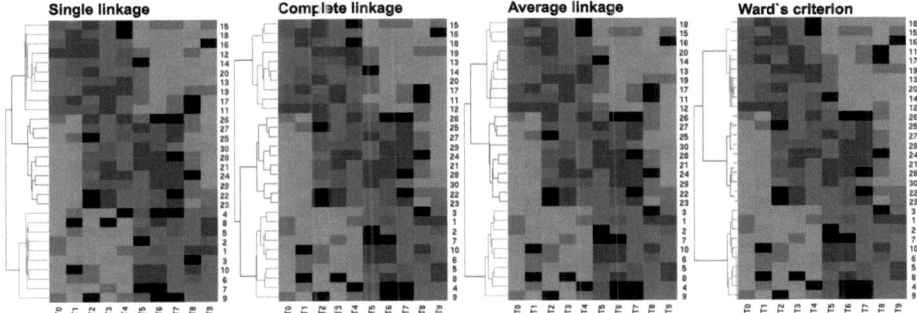

Figure 1.3.: Hierarchical clustering of toy gene expression data. Relative expression levels are color coded; red indicates positive and green negative values. Distances were measured using four different similarity criterions: single-, complete-, average linkage and Ward's criterion. Depending on the criterion, the four resulting trees show different topologies. The colored figure can be found at: http://www.helmholtz-muenchen.de/fileadmin/CMB/IMG/ImgDissLutter/Figure1-3.pdf

on the regulatory impact a particular gene has. Hence, several potentially interesting genes may not be detected within a pairwise comparison analysis.

1.3.1.2. Hierarchical clustering

A somewhat related approach to the detection of differentially expressed genes is the identification of similarities in gene expression patterns. However, unlike comparing the expression of a single gene in different conditions, one here compares the expression patterns of multiple genes with each other. One major goal of this analysis is to identify genes with positively or negatively correlated expression patterns. Genes with a positive correlation in depending on different conditions therefore may also share a common biological function or even are commonly regulated. In contrast, negative correlation of two or more expressed genes may indicate for more or less antagonistic functions.

A common approach to identify correlated genes is clustering. As clustering one denotes the assignment of objects into groups (called clusters) depending on a similarity measure. The objects assigned to the resulting clusters are more similar to each other than objects from different clusters. Similarity is often assessed according to several distance measures, such as euclidean distance or Pearson correlation [Sturn et al., 2002].

Although a bunch of different clustering algorithms exist, in the field of microarray data analysis the most commonly used method is *hierarchical clustering* [Quackenbush,

2001]. The algorithm iteratively connects genes accordingly to their similarity, beginning with the most similar ones. The result is a tree or dendrogram where the branches connect the grouped genes. Cutting the tree at a predefined threshold will give a clustering at the selected precision. Beyond the choice of an appropriate distance measure between distinct genes, the similarity between groups has to be defined, also. Usually the similarity between two clusters can be determined as:

- **Single linkage** or nearest neighbour method. The distance between two clusters i and j is defined as the minimum distance between the elements of each cluster.

- **Complete linkage** or maximum neighbour method. The distance between two clusters i and j is calculated as the maximum distance between an element of cluster i and an element of cluster j.

- **Average linkage** unweighted pair group method (UPGMA). The distance between two clusters is calculated based on the average values using all elements of each cluster.

- **Ward's criterion**. At each step in the analysis, the union of every possible cluster i and j is considered and the two clusters whose fusion results in minimum increase in 'information loss' are combined. Information loss is defined by Ward in terms of an error sum-of-squares criterion, ESS.

Although the algorithm is easy to understand and the results are intuitively interpretable, it also lacks several issues. Depending on the height of the cut of the tree, the size and number of distinct clusters varies. Defining the height that results in the most relevant clusters can not be easily determined. Furthermore, depending on the used distance metric or linkage method, the resulting dendrograms vary (see figure 1.3). Hence, the interpretation of the different results may be misleading or even false. The strength of his method is the unsupervised identification of interesting gene expression patterns. A huge gene cluster showing a distinct pattern can provide novel biological information about regulatory mechanisms. By contrast, a single gene of potential interest may not be identified since it is not assigned to a conspicuous cluster (see chapter 4).

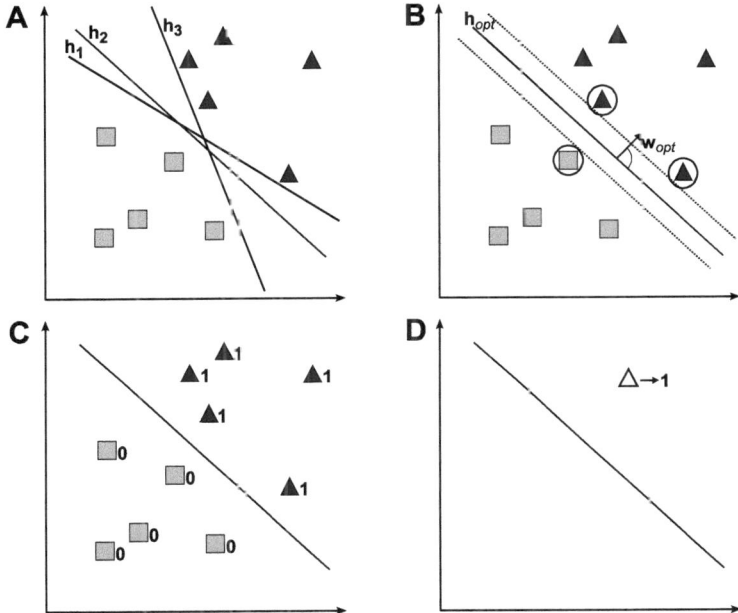

Figure 1.4.: SVM classifier. **(A)** Binary classification. The data is linerly separable by infinite hyperplanes, e.g. $\mathbf{h}_1 \ldots \mathbf{h}_3$. **(B)** A SVM finds the optimal hyperplane $\mathbf{h}_{opt}$ with its normal vector $\mathbf{w}_{opt}$ and the maximum distance to the support vectors (circles). **(C)** The SVM is trained using a training data set. **(D)** A new object can the be classified.

1.3.1.3. Support vector machines

Beyond the identification of strongly differentially expressed genes or genes with common regulatory patterns, one can also try to identify genes, that allow for classification of the dataset. An appropriate and widely used method for this gene selection task is the application of a *support vector machine* (SVM) [Schachtner et al., 2007a; Herold et al., 2008]. This supervised learning approach estimates an optimal hyperplane h which can be characterized by its normal vector $\mathbf{w}$ and a constant b. After training using a finite set of training data, the hyperplane separates the input data into two classes.

The SVM mechanism can be easily illustrated using geometric considerations in a vector space. The training dataset consists of K gene expression profiles. Each gene expression profile is represented by a vector formed by N gene expression values, labeled by two classes. Based on the data, an optimal hyperplane is estimated, that has the maximum possible distance to the training vectors (support vectors) closest to it (see figure 1.4B), and is then characterized by its normal vector $\mathbf{w}_{opt}$. After estimating the optimal hyperplane a new vector $\mathbf{x}$ can be classified according to the decision function (see figures 1.4C,D)

$$f(\mathbf{x}) = \text{sgn}(\langle \mathbf{x}, \mathbf{w} \rangle + b), \tag{1.3}$$

where

$$\mathbf{w} = \sum_{m \in SV} y_m \alpha_m \mathbf{x}_k^{SV} \tag{1.4}$$

and y_m represents the class label, α_m represents a hyperparameter and $\mathbf{x}_k^{SV}$ indicates the support vectors closest to the separating hyperplane. The components of $\mathbf{w}_{opt}$ indicate the importance of a gene for the classification task. Genes with small components in $\mathbf{w}_{opt}$ can be removed as their associated unit vector lies almost parallel to the hyperplane and therefore orthogonal to the optimal class discrimination. Hence, in reverse one can now identify a minimum number of genes, that allow for correct classification. These selected genes may then be used as so-called marker genes, for instance in clinical approaches like cancer classification.

In some cases it might be the case that the data is not linearly separable. In these cases, one can either use soft margin hyperplanes, which allow for some few points to be wrongly classified, or non-linear SVM, where the data is projected into a higher dimensional space using a 'kernel' before classification [Scholkopf and Smola, 2002].

Similar to the pairwise comparison methods, SVMs are based on the power of single gene statistics. Thus, the quality of the trained classifier depends on proper gene expression value normalization. Another problem that may occur is overfitting, especially when the number of features (genes in this case) is large compared to the number of training samples. Unfortunately this is mostly the case in microarray data analysis. To avoid overfitting a preselection of genes, based on gene ranking using pairwise comparison methods, can be applied. Furthermore, in principle SVM are only able to be trained on two different classes.

However, apart from these more technical issues, one emerging problem of SVMs is the potentially misleading interpretation of the selected genes. Genes, that are used to correctly classify the data, are not necessarily genes, strongly related to the conditions under study and, by contrast, genes with a major biological role may not be applicable

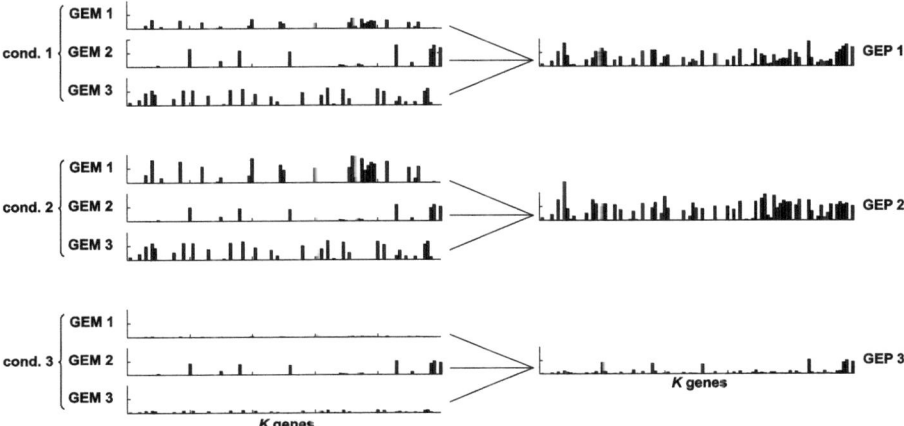

Figure 1.5.: Illustration of the mixing model. K genes differentially contribute to three independent GEMs. Dependent on three different conditions the GEMs are more or less active and superimpose to the three measured GEPs.

for classification.

1.3.2. Mixture models

The basic assumption in the previously discussed mapping models is that the change in gene expression – more or less – directly corresponds to the different conditions. However, according to our knowledge, one gene can be associated with several functions. Thus, a single gene can produce different splice forms with corresponding proteins related to specific tissues or functions [Holmberg et al., 2000; Ryan et al., 2005] and further on, a distinct protein can be part of several pathways or biological processes at once [Alberts et al., 2002]. The composition of the transcriptome within a living cell is controlled by a couple of biological processes, each of which causing its own specific gene expression pattern, the so called *gene expression mode* (GEM). Hence, the expression of a single gene may then be a result of more then one regulatory mechanism. Therefore, we consider a GEP $\mathbf{x_n} = (x_{n1}, \ldots, x_{nK})$, $n = 1 \ldots N$, as the expression level of K genes measured under N conditions resulting in a expression matrix $\mathbf{X} = (\mathbf{x_1}, \ldots, \mathbf{x_N})$, where the columns are formed by the GEPs and the rows correspond to the expression patterns of the distinct genes.

According to this conception, a specific gene expression profile, measured at a distinct condition is then the superposition of simultaneously running processes, each represented by its own GEM. The goal of the following methods is the reconstruction of these GEMs. Unfortunately, the number and properties of the underlying processes are unknown and therefore, the number of possible solutions is infinite. Hence, one has to impose additional restrictions to the model. In general this problem can be specified as a *blind source separation* (BSS) problem, where one tries to recover signals from several observed linear mixtures. In our case mixtures refer to microarray measurements. The following methods are based on decorrelation, independence or non-negativity of the unknown source GEMs. All these methods were developed in the field of linear algebra and are also successfully applied to other BSS problems like removing water artefact's from NMR spectra or functional RMI data analysis [Stadlthanner et al., 2003b; Theis et al., 2005; Böhm et al., 2006].

1.3.2.1. Principal component analysis

One possible approach is to assume that the underlying GEMs forming a GEP are decorrelated. Correlation is a basic statistical measure indicating the strength and direction of a linear relationship between two random variables. *Principle component analysis* (PCA) is a widely used method that allows for the decomposition of several possibly correlated signals into an equal or smaller number of uncorrelated variables. Mathematically speaking, a PCA is a linear transformation that projects multivariate data into a new orthogonal feature space where the first principal component (PC) refers to the direction with the greatest variance and lies on the first new coordinate [Hyvärinen et al., 2001].

Given our data matrix $\mathbf{X}$ where the columns represent the GEPs measured in a microarray experiment and the rows are formed by the single gene expression patterns. PCA now finds an orthogonal transformation $\mathbf{U}$ such that

$$\mathbf{Y^T} = \mathbf{X^T U} = \mathbf{V\Sigma}. \tag{1.5}$$

The columns of the matrix $\mathbf{Y}$ are the principal components, and the columns of $\mathbf{U}$ form the set of orthonormal basis vectors of the PCs. The matrix $\mathbf{\Sigma}$ is a diagonal matrix containing the singular values of $\mathbf{X}$.

As PCA extracts and sorts the PCs according to their variance in decreasing order, a common application is dimensionality reduction. Given the noise present in real data, one can concentrate on the first l components assuming they contain almost all relevant information. In practice a reasonable determination of l is problematic since the amount of noise is generally unknown and the number of components required for a sufficient biological interpretation is hard to define.

However, the application of PCA as a preprocessing step for clustering, compared to clustering of the original data does not necessarily improve cluster quality [Yeung and Ruzzo, 2001]. Since, in this work PCA is only applied as a necessary preprocessing step for independent component analysis (see next section), we here will refrain from a more detailed discussion of PCA.

1.3.2.2. Independent component analysis

The power of PCA is restricted to second order statistics. *Independent component analysis* (ICA) uses the much richer requirement of statistical independence to decompose a given set of measurements into independent source signals so-called independent components (ICs) [Theis, 2002]. To solve this problem, several ICA algorithms have been developed. In this work the two well-established algorithms, JADE [Cardoso et al., 1993; Cardoso and Souloumiac, 1996] and FastICA [Hyvärinen, 1999], implemented in MATLAB® [Mathworks, 2008] were used.

Applied to the analysis of large scale gene expression data, several model assumptions have to be made [Lutter et al., 2008, 2009]. Briefly summarized, gene expression of K genes in a living cell is controlled by M independent biological processes running in parallel. Each process $m \in \{1, \ldots, M\}$ forms a distinct GEM represented by a row vector of K gene expression levels $\mathbf{s_m} = (s_{m1}, \ldots, s_{mK})$. Note that one gene can be part of more than one process/GEM. The respective GEMs superimpose to a measureable GEP (columns of our data matrix $\mathbf{X}$). Although, from our comprehension of the biology of a living cell, no single process is completely isolated, and therefore all processes somehow interact between each other. However, due to a certain autonomy of these processes one can assume that the corresponding GEMs appear to be independent, to a first approximation. ICA decomposes our data $\mathbf{X}$ into a matrix of M independent expression modes $\mathbf{S} = (\mathbf{s_1}, \ldots, \mathbf{s_M})$ and the corresponding $N \times M$ mixing matrix $\mathbf{A}$

including the basis vectors of our new feature space, which then reads as

$$\mathbf{X^T} = \mathbf{AS}. \tag{1.6}$$

Each microarray expression measurement $\mathbf{x}_n$ (columns of $\mathbf{X}$) results from a weighted superposition of independent biological processes. The mixing matrix $\mathbf{A}$ defines the weights with which the corresponding GEM contributes to the measurements or GEPs.

In practice, statistical independence can not directly be determined and therefore, has to be approximated. A common approach to solve this is to approximate independence by non-gaussianity. Non-gaussianity again can be measured by the fourth-order cumulant, the kurtosis. A second measure of non-gaussianity is given by negentropy, which is based on the information-theoretic quantity of entropy [Hyvärinen et al., 2001]. Although further approximations exist, the algorithms used in this work are either based on the kurtosis (JADE) or approximate non-gaussianity using negentopy (FastICA). Furthermore, the reconstruction of independent source signals due to a linear mixture model is limited to two ambiguities [Hyvärinen et al., 2001]:

1. The energy of the variances of the independent signals can not be determined.

2. The order of the reconstructed independent components can not be determined.

Since microarray technology is only capable to measure relative gene expressions (see section 1.2), the first ambiguity is primarily extraneous here. Note that this still leaves the indeterminacy of the sign of the components. Typically, as a result from an ICA one obtains ICs with positive and negative entries, but negative gene expression does not exist. The negative expressions may be considered as related to strongly repressed processes. But, since the sign is unknown, from our gene expression mixture model, it is — without using additional knowledge — impossible to determine whether a strong reconstructed signal corresponds to an activated or repressed biological process.

The second ambiguity is almost equally negligible since we cannot assume that there is any order of the biological processes. However, more relevant for a meaningful interpretation of the results is the relation of a GEM to a specific experimental condition. In case of a time course experiment, the temporal activity of a particular process gives insight into the inner organization of a cell. For instance, in [Lutter et al., 2009] it is shown how a time dependent cellular response to bacterial infection could be reconstructed from determining the contributions of the GEM to the GEP from the mixing matrix **A**.

A further limitation to ICA is the indeterminacy in the overcomplete case. This means, that a unique reconstruction of independent components can only be assured if the number of reconstructed signals is less or equal to the number of used mixtures [Theis and Lang, 2002; Theis et al., 2004a]. Unfortunately, the number of cellular processes is generally unknown. One reconstructed GEM may therefore still represent a superposition of underlying processes. Using a bootstrapping approach, it could be shown that sampled reconstruction is more robust compared to a random model [Lutter et al., 2009] and the results may therefore be interpreted as GEMs referring to single or superpositions of strongly related processes.

1.3.2.3. Non-negative matrix factorization

As a result of an ICA analysis one obtains independent source signals with positive and negative entries. As mentioned the sign of these signals is undetermined, provoking the discussed issue. The Non-negative Matrix Factorization (NMF) techniques replace the assumption of statistical independence by a positivity constraint concerning the entries of the matrices into which the measured GEPs are decomposed [Schachtner et al., 2008]. This constraint of non-negativity seems to be more adequate to microarray data, since gene expressions are measured by strictly positive fluorescence intensities. Applied to our data matrix **X**, where each column represents a GEP and each row a gene expression pattern, NMF approximately factorizes a matrix **X** into a product of two non-negative matrices $\mathbf{W}(K \times L)$ and $\mathbf{H}(L \times N)$ such that

$$\mathbf{X} \approx \mathbf{WH} \qquad (1.7)$$

where the common approach is to minimize $||\mathbf{X} - \mathbf{WH}||$. The columns of **W** are called *metagenes*, while rows of **H** constitute *meta experiments* [Brunet et al., 2004], where L is an integer parameter to be set. In analogy to the ICA mixing model, the metagenes can be interpreted as a particular gene expression mode that is characteristic for a specific biological process. The meta experiments contain the mixing coefficients defining the contributions of each metagene to the experiments. The results can be used to search for potentially interesting source signals, which help to identify putative marker genes

[Schachtner et al., 2007b]. For instance one can search for a specific pattern within the meta experiments to focus only on the genes contained in the corresponding metagene.

However, it has been shown that non-negativity being the only constraint does not lead to unique results [Lee and Seung, 1999, 2001]. As mentioned above, since the number of underlying processes is not known, some more flexibility concerning the number of estimateable sources would be of advantage. By varying the number of estimated sources combined with extensive sampling reproduceable results can be achieved. But in comparison to other well-founded methods, this method still holds the drawback of manual thresholding [Schachtner et al., 2008]. However, in analogy with the ICA model, the number of biological processes to be identified is unclear and, thus one may obtain either superpositions or partially fragmented reconstructed GEMs.

One further solution to increase robustness, is to extend NMF algorithms by additional constraints. In case of reconstruction of GEMs due to biological processes, it is assumed that these processes only correspond to the expression of a few genes, compared to a complete GEP measured with microarrays. Hence, a sparseness measure can be proposed as most appropriate to suitably transform gene expression profiles into interpretable underlying biological signals. Several algorithms applying additional sparseness constraints have been proposed, but either still do not deliver unique results or are extremely computationally extensive [Li et al., 2001; Stadlthanner et al., 2007].

1.4. Conclusions

In this last chapter the basic principles of gene expressionwere outlined and the widely used method of microarray technology was introduced. We then discussed various analysis methods based on two different underlying models towards their biological relevance. In most cases the outcome of these methods is a list of genes of interest. To further interpret these gene lists, a common approach is to create graphs where the nodes represent genes and the edges between the nodes represent distinct relationships between two genes. These relationships can be defined in various ways. For instance one can use gene annotation as provided by the Gene Ontology [Ashburner et al., 2000] or the MeSH database [Nelson et al., 2004]. Another reasonable approach is to use text mining tools as provided by Genomatix® [Genomatix, 2009] or protein-protein interaction networks. The advantages of the representation of genes in a network a diverse. On the one hand a network presentation allows for a more intuitive interpretation of results, where genes related to a specific pathway or involved in a common biological process become easily visible. On the other hand graphs allow for additional analysis methods based on graph theory. These methods can be used to identify interesting network structures or over-represented motifs. In the following network representation will be repeatedly used to illustrate and to interpret the results.

2. Analyzing M-CSF dependent monocyte/macrophage differentiation: expression modes and meta-modes derived from an independent component analysis

2.1. Background

Since microarray technology has become one of the most popular approaches in the field of gene expression analysis, numerous statistical methods have been used to provide insights into the biological mechanisms of gene expression regulation. The high dimension of expression data and the complexity of the regulatory mechanisms leading to transcriptional networks still forces statisticians and bioinformaticians to examine available methods and to develop new sophisticated approaches. However, there are already appropriate methods using different approaches to examine the underlying biological mechanisms determining the gene expression signatures and profiles measured by microarray experiments. Supervised methods using prior knowledge like *Gene Set Enrichment Analysis* (GSEA) deliver useful results under certain conditions. But there is still a lack of reliable data needed for non-classical analysis. Widely used unsupervised approaches, like hierarchical clustering and k-means clustering, use correlations or other distance or similarity measures to identify genes with similar behavior under similar conditions. But these methods are not able to represent more complex structures and interdependencies in the regulatory machinery.

In contrast to the algorithms mentioned above, independent component analysis

(ICA) explores higher-order statistics to decompose observed gene expression signatures (GES), which form the rows of the input data matrix, into statistically independent gene expression modes (GEM), which form the rows of matrix $\mathbf{S}$ according to the data model $\mathbf{X}^T = \mathbf{AS}$. ICA solves blind source separation (BSS) problems, where it is known that the observed data set represents a linear superposition of underlying independent source signals. But it can more generally be considered a matrix decomposition technique which extracts informative features from multivariate data sets like, for example, biomedical signals like EEG (Electroencephalography) [Habl et al., 2000], MEG (Magnetoencephalography) [Vigario et al., 1997] and fMRI (functional magnetic resonance imaging) [Yang and Rajapakse, 2004; Keck et al., 2004; Theis et al., 2004b] recordings. ICA can also be considered a projective subspace technique appropriate for noise reduction [Tomé et al., 2004; Gruber et al., 2006], or artifact removal [Stadlthanner et al., 2003a, 2005] if generated from independent sources.

In this work we will concentrate on the linear case, in which each single microarray GES is considered a linear superposition of unknown statistically independent GEM. To decompose these mixtures into statistically independent components, ICA algorithms like FastICA or JADE have been used. Typically, these GEMs can be interpreted as being characteristic of ongoing, largely independent biological regulatory processes. The philosophy behind can be expressed as: *co-expression means co-regulation*. But the complexity of gene regulation and the various interactions of cellular processes demands a new interpretation of our ICA-derived components. In the following we use these extracted GEMs to generate *sub-modes*, which may provide biological pathway information. The genes contained in these pathway-associated *sub-modes* can be regarded as more or less self-contained parts of larger regulatory networks, which can be represented by combining these *sub-modes* into *meta-modes* according to the functional role of the associated genes.

Here we used M-CSF dependent in vitro differentiation of human monocytes to macrophages as a model process to demonstrate that ICA is a useful tool to support and extend knowledge-based strategies and to identify complex regulatory networks or novel regulatory candidate genes.

The major known pathways associated to M-CSF receptor dependent signaling [Shi and Simon, 2006; Pixley and Stanley, 2004; Ross and Teitelbaum, 2005] include expansion of the role of the MAP-kinase pathway [Wada and Penninger, 2004; Bogoyevitch et al., 2004] and Jun/Fos, Jak/Stat and PI-3 kinase [BehreDagger et al., 1999; Fox et al., 2003; Stephens et al., 2002] dependent signal transduction. Up-regulation of immune-regulatory components involved in innate immunity response (e.g. MHC), specific (e.g. Fcγ) [Houde et al., 2003; Vieira et al., 2002. Booth et al., 2001] and non-specific (CRP, complement, galectins) [Sobota et al., 2005; Swanson and Hoppe, 2004; Mina-Osorio and Ortega, 2004; Lau et al., 2005; Dumic et al., 2006] opsonin receptors as well as charge and motif pattern recognition receptors (e.g. SR-family, LRP, Siglecs etc.)[Fabriek et al., 2005; Minami et al., 2001; Beutler, 2004; Lock et al., 2004], is characteristic for monocyte/macrophage differentiation. Beyond this, an increase of membrane biogenesis, vesicular trafficking and metabolic pathways including amino acids, glucose, fatty acids and sterols, as well as increased activity of lysosomal hydrolases that enhance phagocytotic function [Desjardins, 2003; Martin and Parton, 2006], autophagy [Schmitz and Buechler, 2002] and recycling is triggered through M-CSF signaling as a hallmark of innate immunity [Peiser et al., 2002]. These mechanisms are tightly coupled to changes in cytokine/chemokine response [Branton and Kopp, 1999] and red/ox signaling (NOS e.g. NADPH-Oxidase, Glutathione, Thioredoxin, Selenoproteins) that drive chemotaxis migration, inflammation (e.g.NfκB), apoptosis (eg. Caspases, TP53, NfκB, ceramide) and survival [Forman and Torres, 2002; Nordberg and Arner, 2001; Wang et al., 2006a,b; Cathcart, 2004; Kustermans et al., 2005; Østerud and Bjørklid., 2003].

2.2. Results and Discussion

M-CSF dependent monocyte to macrophage differentiation involves the activation and regulation of many different cellular pathways. In this study we used several microarray experiments and combined them to a data set, which we analyzed using the JADE algorithm. The extracted GEMs were labeled from 1 to 14, according to decreasing energy. Note that the extracted GEMs show positive as well as negative components. They are partitioned into a *sub-mode* containing the negative signals only, denoted by $i.1$, and a corresponding *sub-mode* of the positive signals, denoted by $i.2$, respectively. These *sub-modes* were then combined into so-called *meta-modes* according to the following super categories deduced from the MeSH-filter used: *Apoptosis, signal transduction, cell cycle* and *regulatory sequences*, see table 2.1. Sub-classification and mapping to distinct pathways was then performed with the extracted *sub-modes* using the BiblioSphere MeSH- and GeneOntology-filter tools. Note that our method not only takes into account that one gene can be part of more than one pathway, but also that one pathway can be involved in more than one cellular event. This cannot be achieved with classical clustering tools.

Known M-CSF dependent differentiation pathways		Meta-mode (MeSH Terms)	Pathway	Sub-mode	Mapped genes		
					total	MM	PW
Differentiation	PI3Akt	Signal transduction	MAPK	3.2	67	32	12
	JAK/STAT		II & D	12.2	53	40	18
	MEK/ERK		Cell C.	13.2	60	26	-
				6.2	62	29	29
		Regulatory sequences	JUN/FOS	4.1	43	22	3
			FAM	10.1	48	22	10
				11.2	47	23	12
			TP53 (DNA-P.)	14.1	64	34	12
	TP53 DNA protection	Differentiation cell cycle	TP53 (DNA-protection)	5.2	43	25	8
				11.1	55	22	13
				12.1	64	23	12
Apoptosis			TP53 (DNA-protection)	2.1	55	17	5
				3.1	57	18	6
				6.1	74	28	12
				8.1	37	10	7
		Survival Apoptosis		9.2	43	19	8
	NF-κB			13.1	38	23	7
			BAX	3.1	57	18	10
				8.1	37	10	11
Survival	Prenylated-proteins		CALR	13.1	38	23	16
	Rho kinase		FAS	4.2	58	24	6
				9.2	43	19	11

Table 2.1.: The table shows a comparison of the known M-CSF dependent macrophage differentiation pathways and the results of our gene expression mode analysis as described in the text. MM = meta-mode, PW = pathway, II & D = Innate immunity and defence, Cell C. = Cell communication, FAM = Fatty acid metabolism, DNA-P = DNA-protection.

2.2.1. Signal Transduction

Within the *meta-mode Signal transduction* four *sub-modes*, 3.2, 6.2, 12.2 and 13.2 were combined together. The MAP-kinase pathway (figure 2.1) could be identified as the major signal transduction pathway in *sub-modes* 3.2 and 12.2. *Sub-mode* 6.2 encompassed the functions signal transduction and cell communication. The remaining *sub-mode* 13.2 could not be mapped to a defined pathway, but the majority of genes within this *sub-mode* are associated with innate immunity and defense functions. Among these we identified relevant genes, also related to signal transduction, like CD86, BLNK. The transcription factors LMO2 and FLI1 were unique in *sub-mode* 13.2 whereas MMP9, CD36, CTSK, C1QR1 and MYCL1 as a TF were also present in several other *sub-modes*.

The 12 and 18 respectively, identified MAPK-pathway genes were all unique within their *sub-modes* (table 2.1), except IL8 and DUSP1, which were present in both *sub-modes*. IL8 is a member of the CXC chemokine family and thus one of the major mediators of the inflammatory response. It is also a potent angiogenic factor and has a signalling function in the FAS-pathway, whereas DUSP1 is assumed to play an important role in the human cellular response to environmental stress, as well as in the negative regulation of cellular proliferation. Another central gene of the MAPK-pathway is caspase-1 (CASP1), which was represented in *sub-mode* 12.2 (figure 2.2). Caspase-1 is responsible for the maturation of the multi-functional cytokine interleukin-1β and as member of the FAS caspase cascade it is involved in FAS mediated cell death [Park et al., 2003]. Further remarkable genes associated with MAP-kinase in this *sub-mode* were S100A8, S100A9, GADD45B, CTSK, SOD2 and the transcription factors JUNB and ATF3, since they were all represented in other *sub-modes* or pathways, or play a central role in the MAPK-pathway.

Sub-mode 3.2 combined the MAPK-pathway with the thioredoxin (TRX) reductase/thioredoxin system. TRX is involved in a variety of oxidation reduction reactions that regulate cell growth and survival decisions [Bishopric and Webster, 2002]. It reduces ligand binding and DNA interaction by oxidizing cysteine residues within the DNA binding domain of glucocorticoid hormone receptors. Furthermore, TXNDC14 and TXNRD1 were found in this *sub-mode*. TRX also seems to be up-regulated by NGF through MAPK1 [Masutani H, 2004]. Other genes associated with the MAPK-pathway were: STK17A, SH3BP5, RPS6KA1, CD44, G6PD, IL1RN and the transcription factors EGR2.

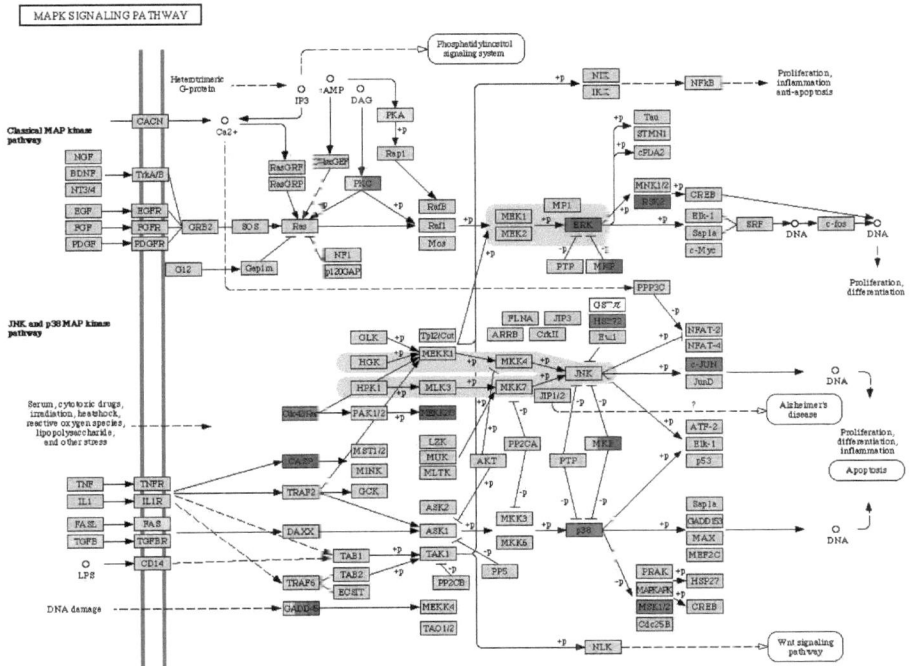

Figure 2.1.: MAP kinase pathway analysis of the *meta-modes*. Yellow boxes correspond to genes mapped to the *apoptosis meta-mode*, red boxes to *regulatory sequences* and blue to *signal transduction meta-mode*, respectively. Solid arrows indicate direct and dashed arrows indirect activation. (Detailed legend information can be found on the KEGG website [Kanehisa et al., 2008]. The colored figure can be found at: http //www.helmholtz-muenchen.de/fileadmin/CMB/IMG/ImgDissLutter/Figure2-1.pdf

In *sub-mode* 6.2 all of the 29 genes involved in signal transduction were also related to the MeSH-term *cell communication*. Five of those signalling genes CFLAR, TXNDC1, YWHAZ, NOTCH2 and PSEN1 were also involved in the negative regulation of cell death.

2.2.2. Regulatory Sequences

The MeSH-term *regulatory sequences* is described as nucleic acid sequences involved in gene expression regulation. This *meta-mode* combines genes mapped to the TP53-

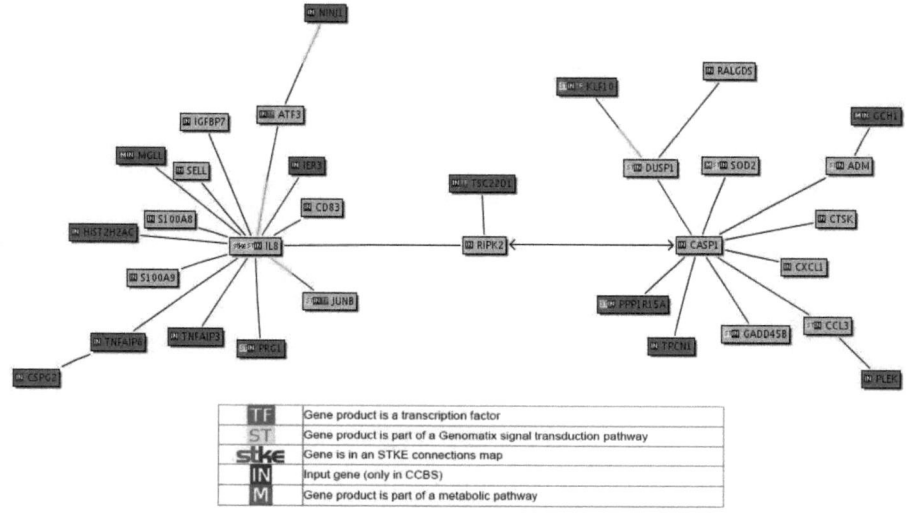

Figure 2.2.: BiblioSphere pathway view shows the mapped Genes of *sub-mode* 12.2. Genes passed the MAPK filter are highlighted blue. Cited relationships between two genes make up the edges. Display of edges is restricted to those that constitute the shortest path from the central node. If a gene that codes for a transcription factor is connected to a gene that is known to contain a binding site for this transcription factor in its promoter, the connecting line is colored green over half of its length near the gene containing the binding site. Arrowheads at the ends of a connecting line symbolize that gene X regulates gene Y. The colored figure can be found at: http://www.helmholtz-muenchen.de/fileadmin/CMB/IMG/ImgDissLutter/Figure2-2.pdf

pathway (*sub-mode* 14.1) and genes related to the oncogenes JUN/FOS (sub-mode 4.1 and 10.1), which are members of a family of transcription factors containing the basic-region-leucine zipper or bZIP motif. The BiblioSphere software did not define a specific pathway for *sub-mode* 11.2, but there were a couple of peptidases and proteinases like LYZ, GGH and CPM as well as a remarkable number of classical targets for the SREBP transcription factors, regulating cholesterol and fatty acid metabolism: SQLE, CYP51A1, HMGCR, FDFT1, INSIG1, IDI1, SC5DL and LDLR.

Sub-mode 14.1 represented an intersection of genes involved in gene expression regulation and the TP53 pathway. Genes which fulfill both criteria were ADM, CCND2, CD59, CDC42, DUSP6, GADD45A, GCH1, IER3, NDUFV2, PIM1, SLC2A3 and UBE3A.

Moreover, *sub-mode* 14.1 received high significance values (Z-Score) for the three other *meta-mode* categories and was also the *sub-mode* with the highest amount of genes represented in other *sub-modes* as well. This can be interpreted as an evidence for the complex and networked nature of gene expression regulation and the interactivity of cellular pathways.

The transcription factor JUN also known as c-Jun belongs to the family of c-Jun N-terminal kinases (JNKs) which are important for development and survival of macrophages [Himes et al., 2006]. *Sub-modes* 4.1 and 10.1 combined twelve genes with a known relationship to the JUN/FOS pathway: CCND2, CREM, CXCL1, GADD45A, IL1RN, JUN, MAPK13, MARCKS, RALA, PLAU, S100A8 and SOD2.

2.2.3. Differentiation, Cell Cycle

The *meta-mode cell cycle* was completely governed by the TP53 pathway. Although all three *sub-modes* 5.2, 11.1 and 12.1 represented TP53 related genes, the intersection of genes was marginal. Only the genes DUSP6, PCNA and PRKCA were mapped to the TP53 pathway and were also present in the *sub-modes* 5.2 and 12.1. *Sub-mode* 11.1 represented genes specialized in cell cycle pathways regulating the interphase and in particular the G1 phase, since it contained the genes PPP1R15A, DUT, CD44, CDKN1A and SMC4L1. *Sub-modes* 5.2 and 12.1 mainly represented genes involved in cell growth and proliferation.

Sub-mode 5.2 was characterized by the TP53 related genes DHFR, VCAN, APP, EIF2AK2 and the transcription factor HMGB2 and HMGB3. Here, the latter has not been mapped to TP53 pathway but is mentioned here because of its strong relation to HMGB2.

The unique TP53 genes in *sub-mode* 12.1 were: CAMK1, CTSB, GSTN1, NME1, HMGCR, GSN, CYP51A1 and IL1RN.

2.2.4. Survival/Apoptosis

Apoptosis related pathways play a major role during the differentiation of monocytes to macrophages. Here we introduce the term "*survival/apoptosis*" for the MeSH term apoptosis, because the identified apoptosis pathways here function as survival mechanisms for the differentiating cells. It has been shown, that an absence of M-CSF induces apoptosis in cultivated monocytes [Becker et al., 1987]. Since apoptosis is regulated through many different pathways and regulatory mechanisms, we could identify seven *sub-modes* (2.1, 3.1, 6.1, 8.1, 9.2, 13.1, 4.2) related to apoptosis. These could be classified to four different pathways involved in the regulation of apoptosis: TP53 pathway, BAX pathway, FAS-pathway and calreticulin (CALR) regulated apoptosis. Three of these *sub-modes* represented only one pathway. *Sub-modes* 2.1, 6.1 were mapped to the TP53 pathway and *sub-mode* 4.2 is governed by CALR regulated apoptosis, whereas the others could be mapped to more than one pathway.

Due to the strongly networked nature of biological regulatory mechanisms, a lot of genes involved in more than one pathway can be regarded as connections between those. Toshiyuki and Reed [Toshiyuki and Reed, 1995] showed that the human BAX-gene is directly regulated by TP53 (TP53), whereas BAX is participating in the regulation of endoplasmatic reticulum Ca^{2+} [Scorrano et al., 2003] as well. In this way it acts as a gateway for selected apoptotic signals. This was represented by the *sub-modes* 3.1, 8.1 and 13.1 which could comparably be mapped to the TP53 and BAX pathway. Sub-mode 8.1 here combined the most interesting combination of genes. The genes CCL3, CCND3, PAICS, FYB, AKAB1, IL1RN, CXCL1, MT1A and the TFs EGR2 and ATF3 could be implicated with BAX. These genes overlapped with five of the seven genes mapped to the TP53 pathway: ATF3, BAX, CSPG2, EIF5B and IL1RN. Furthermore, the metallothioneins which are suggested to regulate DNA binding activity of TP53, MT1A, MT1F, MT1B and MT1X were represented in this *sub-mode* [Ostrakhovitch et al., 2006].

The role of CALR as a major Ca^{2+}-binding (storage) protein in the lumen of the endoplasmatic reticulum is well known [Arnaudeau et al., 2002]. Consequently, one might imagine that CALR is involved in the regulation of apoptotic signals. The following genes of *sub-mode* 4.2 are related to CALR: SLC11A1, CD93, PROCR, NME1 and ATP2B1. All of these genes, except ATP2B1, passed the MeSH-filter apoptosis. The link to the TP53 pathway is the transcription factor FOXO1A (also found in *sub-mode* 6.1) and PRKCB1, which is also involved in various other cellular signaling pathways.

The member of the TNF-receptor superfamily FAS plays a central role in the regulation of programmed cell death. *Sub-mode* 9.2 contained eleven genes related to FAS: GSTM1, RALGDS, ALOX5, VCAN, S100A9, S100A8, VIL2, LY75, STAB1, HEBP2 and CD44.

2.2.5. Otherwise Classified

Although not all *sub-modes* could be mapped to specific *meta-modes*, the remaining *sub-modes* still provide useful information. While the genes sorted to *sub-modes* 7.1 and 7.2 deliver no significant pathway information, they share common behavior. Genes of *sub-mode* 7.1 were all down-regulated in macrophages or up-regulated in monocytes, respectively, whereas genes of *sub-mode* 7.2 were up-regulated in macrophages. Among these, known marker-genes for the different cell types could be identified: MNDA, FCN1 and the S100 calcium binding proteins S100A8, S100A9 and S100A12 as monocyte and IGF2R, TSPAN4, MMP9, CTSK, MMD, TNS1 and CALR as macrophage genes.

Furthermore, the *sub-modes* 5.1, 4.1, 8.2 and 14.2 contained Major Histocompatibility Class (MHC) genes. Whereas the *sub-mode* 5.1 genes HLA-A and HLA-C belong to MHC class I, the MHC genes of the three other *sub-modes* belong to MHC class II which are: HLA-DQB1, HLA-DQA1, HLA-DPB1, HLA-DPA1 and HLA-DMB.

2.3. Conclusions

It has been stated [Liebermeister, 2002; Chiappetta et al., 2004] that the use of ICA for the analysis of gene expression data is a promising tool, but there is still a lack of a careful discussion of the results. Here we emphasized the exploration of the biological relevance and obtained a detailed insight into the networked structure of the underlying regulatory mechanisms. Two MAP kinase related pathways could be identified as the main regulatory pathways during differentiation: the classical MAP kinase pathway and the JNK and p38 MAP kinase pathway, see figure 2.1. These results confirm expectations, according to which the MAP kinase pathway is activated by the M-CSF stimulus and functions as the main signal transduction pathway triggering macrophage differentiation and related pathways.

The conspicuous presence of TP53 associated pathways in M-CSF induced monocyte differentiation is associated with a dramatic regulation of cell-cycle and apoptosis related genes. This leads to the assumption that human mononuclear phagocytes, which are considered to be arrested to non-proliferating cells, still preserve proliferative potential [Martinez et al., 2006].

Furthermore, we could show that ICA is able to distinguish between monocytes and macrophages concerning differential gene expression. This helpful attribute can be used to find specific marker genes not only for different cell types as it is shown here, but also for different tissues or normal and tumor cells.

Moreover, we were able to identify different regulatory mechanisms during M-CSF dependent differentiation. Although signal transduction pathways are mainly regulated by protein modifications like phosphorylation or acetylation, genes associated to specific pathways involved in macrophage differentiation could be separated into *sub-modes* only by analyzing gene expression signatures and their related gene expression modes. Furthermore, this analysis could be improved by combining gene expression *sub-modes* extracted from different microarray experiments into informative gene expression *meta-modes*. The results are in full agreement with the experimental literature on M-CSF dependent differentiation [Schmitz and Grandl, 2007] and illustrate the potential power of such information-theory-based, unsupervised and data-driven analysis.

To fully explore the potential of such information-theory-based unsupervised analysis tools and especially to determine the suitability and reliability of ICA for the analysis of microarray datasets, further investigations are needed. The algorithms still suffer from the fact, that the number of estimated independent components, i.e. the extracted gene expression modes, depends on the number of available gene expression signatures and the dimension of the related gene expression profiles. Therefor, the availability of greater datasets should lead to advancements, and as shown here, greater datasets can be obtained by the careful combination of smaller datasets.

2.4. Methods

2.4.1. Dataset

For our analysis we combined the gene-chip results from three different experimental settings. In each experiment human peripheral blood monocytes were isolated from healthy donors (experiment 1 and 2) and from donors with Niemann-Pick type C disease (experiment 3). Monocytes were differentiated to macrophages for 4 days in the presence of M-CSF (50 ng/ml,R&D Systems). Differentiation was confirmed by phase contrast microscopy. Gene expression profiles were determined using Affymetrix HG-U133A (experiment 1 and 2) and HG-U133plus2.0 (experiment 3) GeneChips covering 22215 probe sets and about 18400 transcripts (HG-U133A). Probe sets only covered by

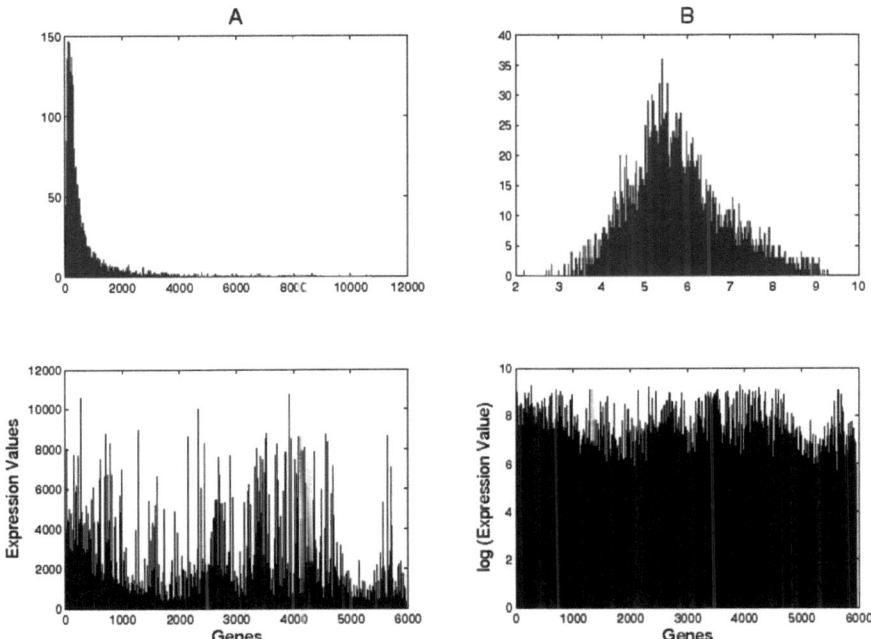

Figure 2.3.: Histograms and expression signatures of an untransformed (A) and logarithmically transformed (B) microarray expression data set.

HG-U133plus2.0 array were excluded from further analysis. In experiment one pooled RNA was used for hybridization, while in experiment two and tree RNA from single donors were used. The final data set consisted of seven monocyte and seven macrophage expression profiles and contained 22215 probe sets. After filtering out probe sets which had at least one absent call, 5969 probe sets remained for further analysis. The complete data set is publicly available in the NCBI Gene Expression Omnibus [Barrett et al., 2007] through the accession number GSE9801.

2.4.2. Preprocessing

The bulk of preprocessing has been done using the Affymetrix GeneChip Operating Software (GCOS), where default presets were used. Additionally, we applied a logarithmic correction to the data. This has been done because effects with multiplicative

55

behavior, which may contain biological relevant information, become linear after logarithmic transformation. Another reason is that, untransformed microarray expression profiles have a strongly skewed, hence unbalanced distribution. This means, that there is a large amount of expression values near zero whereas only very few genes show high expression levels (figure 2.3). To avoid adverse effects caused by such unbalanced distributions, we applied a logarithmic transformation. The final data are usually represented as a data matrix whose columns represent expression signatures of N genes while the rows represent M corresponding gene expression profiles.

2.4.3. JADE-based extraction of gene expression modes

The *Joint Approximative Diagonalization of Eigenmatrices (JADE)* algorithm has been proposed by Cardoso and Souloumiac [Cardoso et al., 1993; Cardoso and Souloumiac, 1996]. It is a nearly exact algebraic approach to perform ICA.

The algorithm JADE is based on fourth-order *cumulant tensors* $\mathbf{T}_z$ of pre-whitened input data $\mathbf{z} = \mathbf{Q}\mathbf{x}$ given by

$$\begin{aligned} Cum(z_i, z_j, z_k, z_l) &= E\{z_i z_j z_k z_l\} - E\{z_i z_j\}E\{z_k z_l\} \\ &\quad - E\{z_i z_k\}E\{z_l z_j\} \\ &\quad - E\{z_i z_l\}E\{z_j z_k\} \end{aligned} \quad (2.1)$$

with the kurtosis $\kappa_i^{(4)} = Cum(z_i z_i z_i z_i)$ being the corresponding autocumulant. Associated with these cumulants is a fourth-order signal space (FOSS) which defines the range of all mappings $T_z : \mathbf{M} \to \mathbf{T}_z(\mathbf{M})$

$$m_{ij} \to [\mathbf{T}_z(\mathbf{M})]_{ij} = \sum_{k,l=0}^{m-1} Cum(z_i, z_j, z_k, z_l) M_{kl} \quad (2.2)$$

The corresponding matrices $[\mathbf{T}_z(\mathbf{M})]_{ij}$ will be called *cumulant matrices* in the following. Note that the dimensionality m of the FOSS equals at most the number of sources.

A spectral representation of the cumulant matrices can be obtained using the column vectors of the whitened mixing matrix with the corresponding eigenvalues related to the kurtosis of the independent components. This spectral representation can be used to obtain an eigenmatrix decomposition of the cumulant tensor according to

$$\mathbf{T}_z(\mathbf{E}^{(q)}) = \mu_q \mathbf{E}^{(q)} \tag{2.3}$$

with $0 \leq q \leq m^2$ symmetric eigenmatrices $\mathbf{E}^{(q)} = \mathbf{u}_q \mathbf{u}_q^T$ and $\mathbf{u}_q$ the q-th column of the mixing matrix $\mathbf{U}$, and μ_q being a scalar eigenvalue. After whitening, a $m \times m$ - dimensional orthogonal matrix $\mathbf{D} = [\mathbf{d}^{(0)} \ldots \mathbf{d}^{(m-1)}]$, which jointly diagonalizes all eigenmatrices of $\mathbf{T}_z$, is found by maximizing the *joint diagonality criterion*

$$c(\mathbf{D}) = \sum_{q=0}^{m^2-1} \left| Diag\left(\mathbf{D}^T \mathbf{E}^{(q)} \mathbf{D}\right) \right|^2 \tag{2.4}$$

where $Diag(.)$ denotes the vector of diagonal matrix elements. The joint diagonalizer $\mathbf{D}$ is then equivalent to the whitened mixing matrix $\mathbf{U}$, hence the unknown independent component expression mode can be estimated easily.

2.4.4. Sub-modes and meta-modes

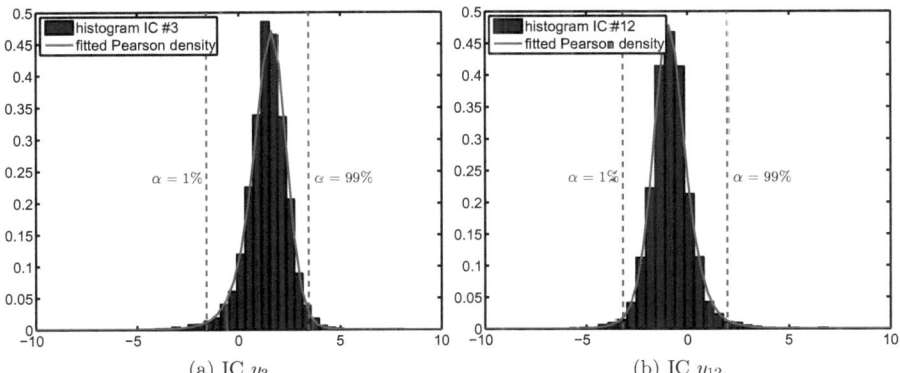

Figure 2.4.: Maximum-likelihood Pearson fit of the EM-densities, for EM number 3 in (a) and number 12 in (b). The corresponding four moments are $\mu(y_3) = 1.4$, $\sigma(y_3) = 1.0$, skewness$(y_3) = -0.95$ and kurtosis$(y_3) = 4.0$ for (a) and $\mu(y_3) = -0.84$, $\sigma(y_3) = 1.0$, skewness$(y_3) = 0.49$ and kurtosis$(y_3) = 4.4$ for (b).

As result of an ICA analysis of a set of gene expression signatures representing the rows of the transpose data matrix $\mathbf{X}^T$, we obtain a matrix $\mathbf{S}$ of independent components (the rows of $\mathbf{S}$) which represent independent gene expression modes (GEMs) as well as a matrix $\mathbf{A}$ of basis vectors of the new feature space. To deduce meaningful biological information from the GEMs, the discovery of specific biological processes, which determine the modes, is the goal of our expression mode analysis. After decomposing the data matrix with ICA, each GEM has been split into two *sub-modes* which can be considered to feature genes which are co-expressed, thus co-regulated by the underlying regulatory process. A GEM consists of scores of gene contributions to the *sub-modes* which account for the observation that excitatory as well as inhibitory regulations exist. In order to extract the most significant genes, various statistical tools can be applied which, however, often suffer from the small M large N case. Therefore, in most cases a threshold is simply applied, or, after ranking, a fixed number of top and bottom genes are chosen and further analyzed [Lee and Batzoglou, 2003]. The rational behind these methods is that each extracted gene expression *sub-mode* is best represented by its most active genes. However, the choice of threshold or number of active genes is non-trivial, and will influence the results considerably. In this study we assume instead that mapping to distinct pathways is most non-ambiguous by using a relatively small number of genes.

Here, we took a different approach by selecting genes that are extremal with respect to some probabilistic model. For each GEM $y(i) \in \mathbb{R}$, where i indexes the genes, we calculated the first four central moments corresponding to mean, standard deviation, skewness and kurtosis of the underlying data distribution. These shape parameters are then used to fit a density according to the Pearson family [Nair and Sankaran, 1991] using maximum-likelihood, see figure 2.4. We chose a Pearson density as prior since it allows for flexible modeling with respect to these first four moments, which seemed crucial as for example skewness varies considerably between modes, see figure 5, and high kurtotic as well as close-to-Gaussian modes were present.

We then used the estimated Pearson densities to determine the $1-\alpha$ and α percentiles for $\alpha = 1\%$. Samples that lie below the 1-percentile are denoted as significantly down-regulated genes, and genes above the 99-percentile as significantly up-regulated genes. The corresponding *sub-modes* were labeled as $i.1$ for down-regulation and $i.2$ for up-regulation. In table 2.2 we list the number of significant genes in each *sub-mode*.

2.4.5. Mode analysis

EM	n_{down}	n_{up}
1	68	112
2	59	59
3	69	79
4	54	64
5	74	47
6	88	65
7	54	68
8	43	64
9	59	51
10	51	38
11	64	59
12	71	62
13	43	73
14	79	34

Table 2.2.: Number of selected down- and up-regulated genes in each gene expression mode (GEM).

We analyzed the gene *sub-modes* with BiblioSphere (http://www.genomatix.de). BiblioSphere is a data mining tool intended to provide gene relationships from literature databases and genome-wide promoter analysis. The probe sets were mapped to transcripts and to known genes with use of the Genomatix database. To uncover the biological meaning of the genes in the *sub-mode*, we applied the MeSH-Filter (Medical Subject Headings) to our data, which is the National Library of Medicine's controlled vocabulary thesaurus. We decided to use the category *biological sciences* as filter criterion. Co-citations between the genes of the *sub-mode* were taken into account by using the literature mining tool of the BiblioSphere software. Interesting terms were identified through Z-Scores which indicate over-representation of genes in the referring biological categories. Z-Scores are given by $Z-Score = (n-\hat{n})/\sigma_n$ where n is the number of observed genes meeting any given criterion, $\hat{n}$ is the corresponding expected number and σ_n gives the standard deviation of n. All terms mentioned in this work are significant with respect to the Genomatix guidelines.

Depending on our filter analysis we defined several *meta-modes*, where we combined *sub-modes* with similar categories. In some cases we subclassified *sub-modes* within one *meta-mode*. In this way 4 meta-modes could be generated, whereas 17 of 28 *sub-modes* could be mapped to at least one *meta-mode*. For some *meta-modes* we displaced the MeSH-Term category with additional categories with respect to the underlying biology.

Additionally we used the KEGG pathway database for biochemical pathway analysis to more thoroughly characterize the biological relevance of a *meta-mode*. The genes corresponding to the *meta-modes* were mapped on database pathways using Pathway-Express which is part of the Onto-Tools provided by Intelligent Systems and Bioinformatics Laboratory [Draghici et al., 2007].

3. Analyzing time-dependent microarray data using independent component analysis derived expression modes from human Macrophages infected with *F. tularensis holartica*

3.1. Introduction

Environmental stimuli or the activity of the internal state of cells induce or repress genes via up- or down-regulation of corresponding expressed mRNAs. Gene expression is controlled by a combination of mechanisms including those involving networks of signalling molecules, transcription factors and their binding sites in the promotor regions of genes, as well as modifications of the chromatin structure and different types of post-transcriptional regulation. The expression of each gene thus relies on the specific processing of a number of regulatory inputs.

High-throughput genome-wide measurements of transcript levels have become available with the recent development of microarray technology [Stekel, 2003]. Intelligent and efficient mathematical and computational analysis tools are needed to read and interpret the information content buried in these large data sets (see section 1.3).

Traditionally two strategies exist to analyze such data sets. If prior knowledge about classification of the samples is available, a *supervised*, also called *knowledge-based*, analysis can identify gene expression patterns, called features, specific to a given class, which can be used to classify new samples. Without any hypothesis, *unsupervised*, i.e. data driven, approaches can discover novel biological mechanisms and reveal genetic regulatory networks in large data sets. Such unsupervised analysis methods for microarray data analysis can be divided into clustering approaches, model-based approaches and projection methods. Clustering approaches group genes by some measure of similarity. A fundamental assumption of such clustering approaches is that genes within a cluster are functionally related. In general, no attempt is made to model the underlying biology. A drawback of such classical methods is that clusters generally are disjunct but genes may be part of several biological processes. Model-based approaches try to explain the interactions among the biological entities with the help of hypothesized concepts. Parameters of the model can be trained from expression data sets [Friedman, 2004]. With complex models not enough data may be available to properly estimate the parameters, hence overfitting may result. Projective subspace methods try to expand the data in a basis with desired properties. Projective subspace methods commonly used are principal component analysis (PCA), independent component analysis (ICA) or non-negative matrix factorization (NMF). Note that often PCA is a necessary preprocessing step for ICA algorithms. Here we focus on the well-known stochastic FastICA algorithm to analyze our time-dependent gene expression profiles (GEPs).

ICA decomposes the GEPs into statistically independent *gene expression modes* (GEM), the so-called independent components (ICs) [Cichocki and Amari, 2002]. The algorithm FastICA assumes a linear superposition of these unknown GEMs, also called source signals, forming the observed GEPs measured with microarray gene chips. Each retrieved GEM is considered to reflect a basic building block of a putative regulatory process, which can be characterized by the functional annotations of the genes that are predominant within the component. Each GEM thus defines corresponding groups of induced and/or repressed genes. Genes can be visualized by projecting them to particular expression modes which help to highlight particular biological functions, to reduce noise, and to compress the data in a biologically meaningful way.

In this work microarray data of human macrophages, deduced from human monocytes by M-CSF triggered differentiation and infected with a *F. tularensis holartica* strain called LVS (live vaccine strain), were analyzed. Our aim was to determine the global gene expression profile of human macrophages from three different donors infected *in vitro* with *F. tularensis* LVS. Expression profiles were followed over a period of 72h, resulting in a series of ten experiments. To monitor assay and hybridization performance, a set of quality parameters (poly-A controls, hybridization controls, percent present, background and noise values, scaling factor) were assessed. None of them exceeded the given ranges, indicating that our data is of high quality. An analysis of these experiments using the FastICA algorithm [Hyvärinen, 1999] is reported in this work.

3.2. Methods

3.2.1. Sample preparation and expression level calculation

Human monocytes were obtained from three healthy donors by diagnostic leukapheresis and counterflow elutriation as described previously [Langmann et al., 2003] under full GLP (good laboratory practice) conditions. The cells were cultured on plastic petri dishes in macrophage SFM medium (Gibco BRL, Karlsruhe) and allowed to differentiate for 5 days in the presence of 50 ng/ml recombinant human M-CSF (R&D Systems, Wiesbaden, Germany) to macrophages. Finally, the cells were infected with *F. tularensis* LVS. Three independent *F. tularensis* LVS infection experiments were chosen for further analysis. The infection rates and the percentage of living cells were comparable in all three experiments.

Total RNA was extracted from cultured cells according to the manufacturer's instructions using the RNeasy Protect Midi Kit (Qiagen, Hilden, Germany). Purity and integrity of the RNA was assessed on the Agilent 2100 bioanalyzer with the RNA 6000 Nano LabChip® reagent set (Agilent Technologies, USA). The RNA was quantified spectrophotometrically and then stored at -80 °C. At each timepoint enough total RNA could be isolated for DNA-microarray analysis and subsequent realtime RT-PCR verification experiments. The quality assessment of RNA samples is a major point in DNA-microarray analysis. All RNAs were of superior quality without any signs of mRNA degradation. The RNA integrity number (RIN) was close to the optimum (10) in all experiments.

Gene expression levels were measured using Affymetrix GeneChip® HGU133 Plus 2.0 Arrays. Array comparison analysis was carried out by calculating expression levels and fold changes using Affymetrix GeneChip Operating Software (GCOS). Expression values after 0.5h, 1h, 2h, 3h, 6h, 9h and 12h of incubation with 100 MOI (multiplicity of infection) *F. tularensis* LVS were compared to the 1h control incubation. Furthermore, infected and control probes were compared after incubation at 24h, 48h and 72h.

3.2.2. Model assumptions

The transcription level of all genes in a cell is the result of the action of several regulatory processes which in parallel control the response of a cell to external stimuli. Matrix decomposition techniques set out to factorize a set of observed GEPs into components according to some specified constraints to assure unique decompositions. Such constraints then lead to either *statistically uncorrelated* (PCA) or even *statistically independent* (ICA) components. The latter may often be identified as regulatory processes governed by signalling pathways which are only weakly coupled to each other and can be considered as acting independently of each other to a first approximation. Each such process can then be represented by a vector of expression levels of up- or down-regulated genes, the *gene expression modes* (GEMs). Under each experimental condition, the different regulatory processes then linearly superimpose the expression levels of each gene according to the different GEMs to result in the observed GEPs measured by a microarray sample. The justification of such simplifying assumptions comes from the "biological meaning" of the resulting expression modes extracted by such matrix decomposition techniques. If such GEMs can clearly be identified with known signalling pathways within a cell for the problem at hand, the model decomposition is justified. Otherwise non-linear decompositions might need to be considered. For such matrix factorization algorithms to be applied, centered data, i.e. $\langle \mathbf{x} \rangle = 0$, will be assumed for simplicity. This can always be achieved by subtracting a time averaged expression level from each data point.

3.2.3. ICA model

Given the state of a cell at the time of experiment is governed by M regulatory processes $\mathbf{S} = (\mathbf{s}_1, \ldots, \mathbf{s}_M)^T$ which are considered reasonably independent of each other and operate in parallel, and where each of them is represented by a row vector of K gene expression levels, i.e. $\mathbf{s}_m = (s_{m1}, \ldots, s_{mK})$, then $\mathbf{S}$ forms a $M \times K$ matrix whose rows consist of statistically independent GEMs. Each such mode forms a component expression pattern or component signature, in which the contribution of each gene to the envisaged independent regulatory processes is reflected via its expression level. Within a microarray experiment, the level of expression of all genes $\mathbf{x}_n = (x_{n1}, ..., x_{nK})$ is measured under N different experimental conditions, resulting in a microarray expression matrix $\mathbf{X} = (\mathbf{x}_1, \ldots, \mathbf{x}_N)^T$, where the rows form the GEPs $\mathbf{x}_n$. Hence, a microarray data matrix $\mathbf{X}$ can be formed with N rows, representing GEPs, and K columns, representing the expression levels of a gene across all experimental conditions. Assuming that different experimental conditions cause different expression levels of each gene within the independent regulatory processes, each observed GEP, i.e. each row of $\mathbf{X}$, results as a weighted superposition of the independent GEMs, represented by the rows of $\mathbf{S}$. In matrix notation this model then reads

$$\mathbf{X} = \mathbf{APDS}, \qquad (3.1)$$

where $\mathbf{A}$ represents the $N \times M$ matrix of mixing coefficients and here we set $N = M$. The *under-determined* or *over-determined* cases with $N \neq M$ is more difficult and will not be considered here. The N columns of $\mathbf{A}$ may be considered to form a new representation with basis vectors $\mathbf{a}_m = (a_{m1}, \ldots, a_{mN})$, also called feature profiles (FP), where each a_{mn} defines the weight with which the nth GEM contributes to the mth observed GEP. In addition, the matrices $\mathbf{P}$ and $\mathbf{D}$ account for trivial permutation and scaling indeterminacies.

By approximating the negentropy as a measure of statistical independence, the FastICA algorithm computes a de-mixing matrix $\mathbf{W}$ such that

$$\mathbf{Y} = \mathbf{WX}, \qquad (3.2)$$

where $\mathbf{Y}$ represents a matrix of transformed variables $\mathbf{y}_1, \ldots, \mathbf{y}_N$, which correspond to the extracted independent components or GEMs subject to scaling ($\mathbf{D}$) and permutation ($\mathbf{P}$) indeterminacies [Comon et al., 1994]. They are extracted from the data by the algorithm as statistically independent as possible, and represent close approximations of the unknown expression signatures of the hypothetical underlying regulatory processes represented by $\mathbf{s}_1, \ldots, \mathbf{s}_N$.

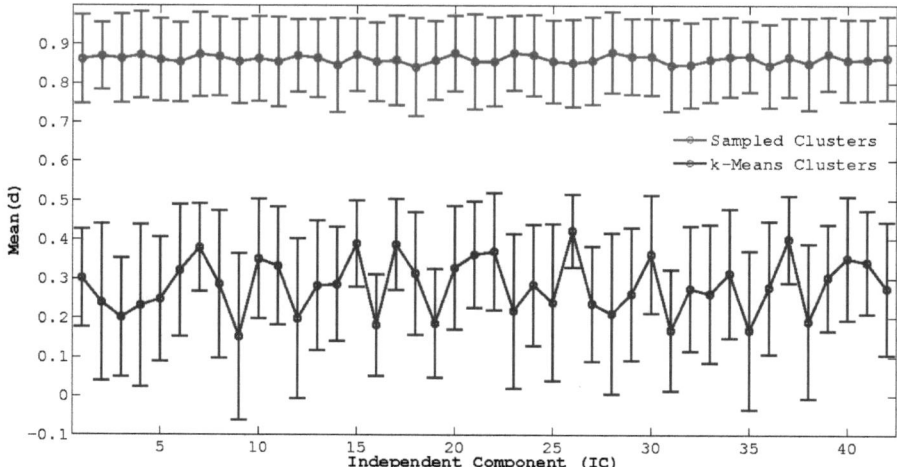

Figure 3.1.: The means and the standard deviations of the differences d of all clustered row vectors $\mathbf{w} \in \mathcal{W}$ to the corresponding code book vector $\mathbf{c}_p$ for each independent component (IC) compared to a null model of randomly sampled clusters. The colored figure can be found at: http://www.helmholtz-muenchen.de/fileadmin/CMB/IMG/ImgDissLutter/Figure3-1.pdf

3.2.4. Stability Analysis

The number of GEMs extracted by the FastICA algorithm corresponds to the number of experiments, i.e. the number of different microarray data sets available. As the number of underlying independent regulatory processes contributing to any observed set of expression signatures is generally unknown, the GEMs extracted, due to the independence constraint enforced by the data matrix decomposition, may, at least to some extent, still represent superpositions of such underlying regulatory processes being searched for. This fact results in fluctuations in the estimated GEM upon repeated decomposition of the given data matrix. Unfortunately, these fluctuations also sometimes confounds the immediate and straightforward biological interpretation of such modes. Despite this it is the hope of every matrix decomposition analysis that the resulting GEMs provide for a more intuitive and insightful interpretation of the observed states of the cell under the experimental conditions and environmental stimuli to which it was exposed.

Because FastICA belongs to the class of stochastic matrix decomposition algorithms, the robustness of its results needs to be assured. To test the robustness of the resulting

GEMs, we performed a bootstrap analysis. To do so, we randomly generated 50 subsamples with a sample size 25% smaller than the original data set. As a consequence, repeating the analysis $L = 50$ times might render some or all of the extracted components to differ slightly in the various repeats. We then estimated the robustness of these repeatedly extracted GEMs.

We combined the rows $\mathbf{w}_n^l$ to a set $\mathcal{W}$ of row vectors, where l represents a particular ICA run and n is the nth row of the de-mixing matrix $\mathbf{W}^l$. Because $\mathbf{W} = \mathbf{A}^{-1}$ each row vector $\mathbf{w}_n$ contains the weights with which each observed GEP is combined to an extracted GEM. Using a projective k-means clustering [Gruber et al., 2006] the resulting row vectors are then clustered into N clusters according to the following metric representing our distance or similarity measure:

$$d(\mathbf{w}, \mathbf{v}) := \sqrt{1 - \left(\frac{\mathbf{w}^T \mathbf{v}}{\sqrt{\|\mathbf{w}\|\|\mathbf{v}\|}}\right)^2} \qquad \mathbf{w}, \mathbf{v} \in \mathcal{W} \qquad (3.3)$$

Now we use the centers of gravity of each cluster as code book vectors $\mathbf{c}_n, n = 1, \ldots, N$ for our stability analysis. The result of the clustering can be described by the sets $\mathcal{W}_n = \{\mathbf{w} \in \mathcal{W} \mid s(\mathbf{w}) = \mathbf{c}_n\}$ with $s(\mathbf{w}) = \arg\min_n d(\mathbf{w}, \mathbf{c}_n)$.

We evaluated the quality of each cluster $\mathcal{W}_n$ by calculating the 1st and 2nd moment of the distance distribution within each cluster, i.e. the empirical mean and standard deviation of all distances between the code book vector $\mathbf{c}_n$ of cluster n and the data vectors within the cluster using the distance measure d as defined above. In particular, $\text{mean}_n = \text{mean}(\{d(\mathbf{w}, \mathbf{c}_n) \mid \mathbf{w} \in \mathcal{W}_n\})$ and $\text{var}_n = \text{var}(\{d(\mathbf{w}, \mathbf{c}_n) \mid \mathbf{w} \in \mathcal{W}_n\})$ (figure 3.1). As a null model we randomly sampled N clusters from $\mathcal{W}$ with size L. For each sampled cluster we calculated the mean and standard deviation of all distances between the sampled vectors and the respective projective centroid.

3.2.5. Grouping genes

Each estimated GEM contains the gene expression levels of all genes within any given microarray experiment, i.e. every experimental condition chosen. Assuming that the genes involved in a hypothetical regulatory process represented by the GEM show relatively high expression within this GEM, then those genes are of utmost interest which correspond to the most or the least expressed. Only genes whose expression level exceeded the mean expression level plus five times the standard deviation of the considered GEM were retained for further analysis. These genes have been grouped together into gene groups of size between 35 and 94 genes, containing the most strongly expressed or suppressed genes. Remember that one gene may be involved in more than one regulatory process, i.e. its expression level may be high or low in several gene expression modes.

3.2.6. Biological relevance

Further information about the biological relevance of the genes and their regulation mechanisms can be gathered from public databases such as *Gene Ontology* (GO) (available at http://www.geneontology.org/). The biological information available within GO can be further explored using software tools like *Onto-Express* [Draghici et al., 2007] (available at http://vortex.cs.wayne.edu/Projects.html) or *Genomatix BiblioSphere* (see http://www.genomatix.de/).

BiblioSphere provides further biological information by structuring input data into biological pathways, i.e. networks of interacting genes thereby delivering systems biology knowledge to organize genes within groups into functional networks. The interaction network is a data-mining solution in which relationships from literature databases, genome-wide promoter analysis and verified gene interactions are combined. Results can be classified by tissue, Gene Ontology and MeSH (see http://www.nlm.nih.gov/mesh/).

Statistical rating by Z-scores indicate over- and under-representation of genes in the certain biological categories which are organized into hierarchies. For each term in the hierarchy, a statistical analysis is performed based on the number of observed and expected annotations. With each associated GO or MeSH term a Z-score is provided measuring the relevance of the functional term within the context of the group of genes under consideration. Z-scores are given by $Z\text{-}score = (n - \hat{n})/\sigma_n$, where n is the number of observed genes meeting any given criterion, $\hat{n}$ is the corresponding expected number and the standard deviation σ_n measures the fluctuations of n around the mean. The Z-score of this term helps to estimate whether a certain annotation, or group of annotations, is over- or under-represented in the tested set. Such score helps to determine whether the accumulation of annotations in a certain branch of the hierarchy is meaningful.

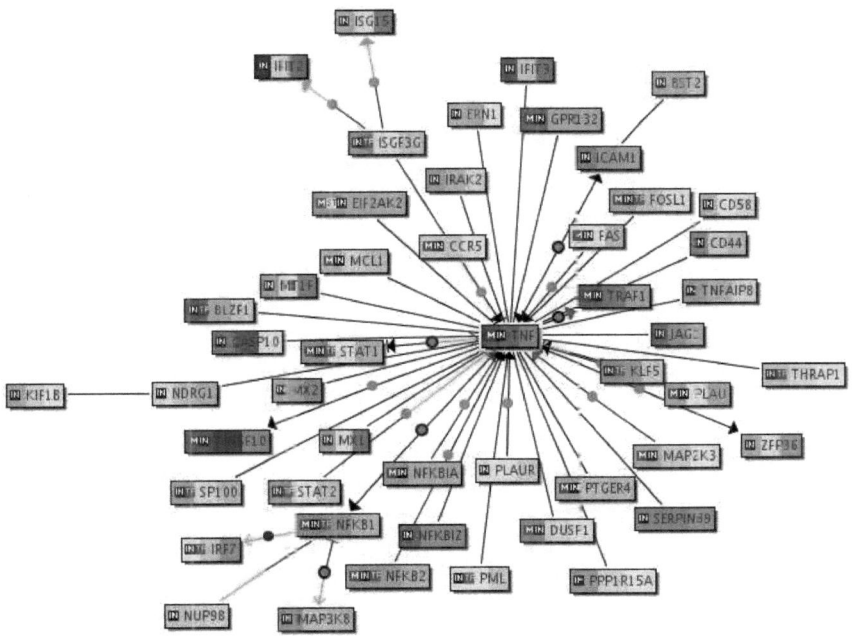

Figure 3.2.: Functional gene networks resulting from a hierarchical clustering analysis. Expression levels for each gene are color-coded. Overexpression is colored red, underexpression blue. The stripes from left to right code for early, middle and late response. Cited relationships between two genes make up the edges. Display of edges is restricted to those that constitute the shortest path from the central node. If a gene coding for a transcription factor is connected to a gene with a predicted binding site in its promoter, the connecting line is colored green over half of its length near the target gene. Arrowheads at the ends of a connecting line symbolize regulation. Hand-annotated gene-gene relationships are indicated by a circle in the center of the connection line. The colored figure can be found at: http://www.helmholtz-muenchen.de/fileadmin/CMB/IMG/ImgDissLutter/Figure3-2.pdf

3.3. Results

3.3.1. Pathways biostatistics

For a knowledge-based pathway analysis, all expressed genes from the three LVS infection experiments were mapped to 78 manually annotated biomedical pathways. To avoid a proband specific bias and to determine a global expression profile, only those genes were retained which displayed similar responses (up-/down-regulation) in all three probands across all measurements. This analysis resulted in 54 genes (52 induced genes, 2 repressed genes) indicating that Chemokine signaling, interleukin 1 and TNF-response as well as NFκB signaling are the major pathways strongly influenced by LVS. Prostaglandin synthase 2 and superoxide dismutase 2 are also induced. Lysophospholipase 3 and zinc finger protein 589 are the only repressed genes detected.

3.3.2. Hierarchical clustering

As a further analysis method, we performed a hierarchical clustering on the data set and selected clusters of differentially expressed genes which show similar time dependent behavior over all three donors. This resulted in 3 clusters corresponding to an early (35 genes), a middle (54 genes) and a late (89 genes) response.

To further define the regulatory network between these genes and to search for interdependent activation waves, Genomatix BiblioSphere analysis was carried out with these data sets. Functional analysis based on the MeSH Filter "Disease" resulted in the following top five terms with good Z-scores for each of the three response terms (table 3.1). To gain a focused view on a disease related network, genes related to the top terms of each cluster were combined. This resulted in a network of 49 genes which was analyzed again using BiblioSphere (figure 3.2). The corresponding regulatory network is centered around TNF. As can be seen, the expression levels of genes encoding TNF, as well as TNF-interacting proteins like (TRAF1, TNFAIP8), adhesion molecules (ICAM1) and kinases increase rapidly and decline at later times thus representing an *early response*. At these early times, signal transducer and activator of transcription genes (STAT1/2) are predominantly weakly expressed. In a second signaling wave, the expression levels of TNF induced genes such as the transcription factor NFκB (NFκB1, NFκB2, NFκBIA) and their target genes (IRF7, NUP98, MAPK3K8) increase during an intermediate time

Resp.	MeSH Term	Z-score	Percent.
ER	Inflammation	53.03	31%
ER	Sepsis	24.32	26%
ER	Systemic Inflammatory Response Syndrome	22.97	26%
ER	Reperfusion Injury	20.86	14%
ER	Shock	18.31	20%
MR	Inflammation	22.6	9%
MR	Cell Transformation, Neoplastic	14.45	17%
MR	Cell Transformation, Viral	10.26	7%
MR	Leukemia-Lymphoma, T-Cell, Acute, HTLV-I-Assoc.	9.56	2%
MR	HTLV-I Infections	8.85	2%
LR	Leukemia, Promyelocytic, Acute	155.37	9%
LR	Leukemia, Nonlymphocytic, Acute	81.32	12%
LR	Leukemia, Myeloid	65.03	15%
LR	Leukemia	52.06	18%
LR	Translocation, Genetic	42.02	7%

Table 3.1.: Terms and Z-scores resulting from a hierarchical clustering and MeSH filtering. ER = early response; MR = middle response; LR = late response. Also the fraction of the genes associated with each MeSH term is given in %

interval representing a *middle response*. During a final *late response*, TNF expression declines and expression of the concomitant signaling genes decreases (NFκB1/2, Rel). Late cytokine response, represented by the interferon-induced proteins (IFI2/3, MX1/2), is continually increased during the kinetic experiment. An overlap between these regulatory models and the top 54 genes from the pathway analysis concerning inflammation associated genes like ICAM1, IRAK2, JAG1, NFKB1, NFKB2, TRAF1 and TNF is observed.

3.3.3. ICA analysis

As a result of the ICA analysis, we obtained $N = M$ expression modes which represent the hypothetical gene regulatory processes. To identify relevant processes represented by the extracted GEMs, we analyzed time dependent patterns formed by the FPs setting up the mixing matrix **A**. To avoid a proband specific bias we filtered out FPs similar among all three probands. Therefore we split up each FP into proband specific temporal patterns and compared them by calculating correlations. Only those FPs which show a high correlation (above 0.8) between all probands specific patterns were used for further analysis. To find FPs comparable to the clusters derived by the hierarchical clustering

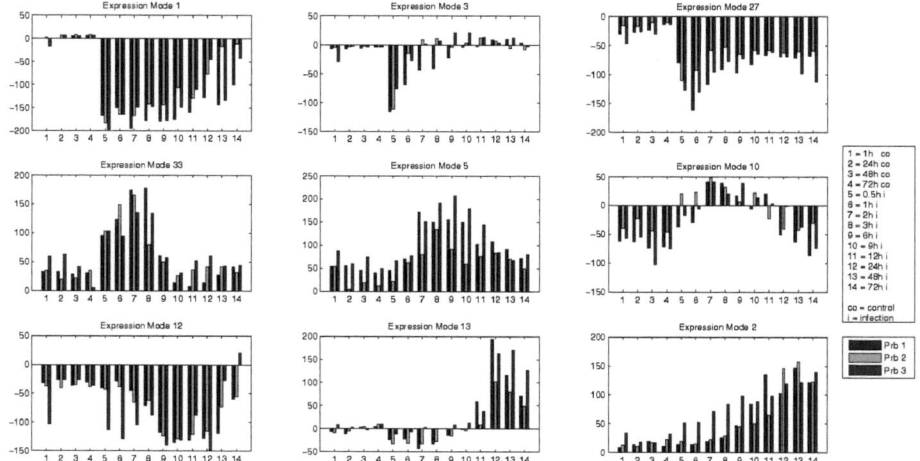

Figure 3.3.: Feature profiles with similar temporal patterns for all three probands (Prb 1-3). Blue, green and red bars. Shown are only those, used for time dependent response analysis: *top*: early response, *middle*: middle response, *bottom*: late response. Gene *response groups* were created from the corresponding gene expression modes. See text for a detailed explanation. The colored figure can be found at: http://www.helmholtz-muenchen.de/fileadmin/CMB/IMG/ImgDissLutter/Figure3-3.pdf

approach, we identified those with temporal patterns showing high early, middle or late response activity (figure 3.3). We have chosen three FPs for each response type respectively, and merged the extracted gene groups from the corresponding GEMs to three *response groups* (RG) called *early* (149 genes), *middle* (171 genes) and *late* (158 genes).

The biological relevance of these RGs was explored using the Genomatix software. We analyzed each RG using the MeSH Filter "Disease". This resulted in a list of the most related MeSH terms (see table 3.2). They are strikingly different to the MeSH terms derived from hierarchical cluster analysis, and in accordance, the ICA derived terms show noticeably higher Z-scores (Inflammation, Systemic Inflammatory Response Syndrome). Furthermore, ICA results show Inflammation as the highest ranked term in all three responses. The percentage of genes associated to MeSH-terms is consistently higher in ICA derived RGs.

The additionally derived network can be seen in figure 3.4. The early response is

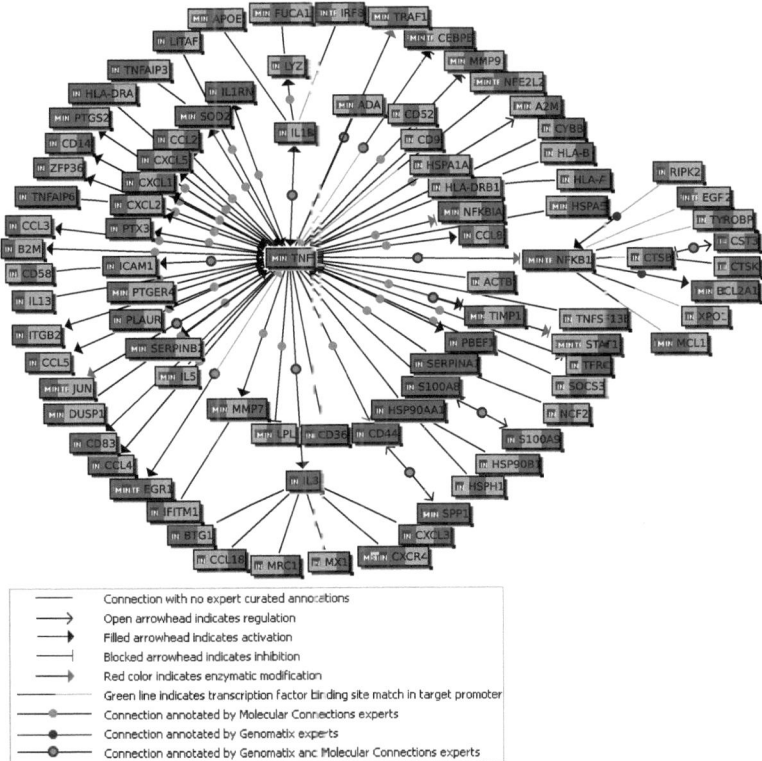

Figure 3.4.: Functional gene network resulting from the ICA analysis. Stripes from left to right code for early, middle and late response group. If a gene is a member of one or more of the response groups the stripe is colored red. Edges between two genes denote co-occurrence within one abstract. Display of edges is restricted to those that constitute the shortest path from the central node. 'TF' stands for transcription factor, 'ST' means gene is part of Genomatix signal transduction pathway, 'IN' means input gene and 'M' marks a gene which is part of a metabolic pathway. The colored figure can be found at: http://www.helmholtz-muenchen.de/fileadmin/CMB/IMG/ImgDissLutter/Figure3-4.pdf

Response	MeSH Term	Z-score	Percentage
ER	Inflammation	93.74	52%
ER	Bacterial Infections and Mycoses	49.36	48%
ER	Arthritis	44.51	40%
ER	Joint Diseases	43.63	40%
ER	Systemic Inflammatory Response Syndrome	42.95	33%
MR	Inflammation	64.35	49%
MR	Bacterial Infections and Mycoses	30.61	40%
MR	Systemic Inflammatory Response Syndrome	27.35	23%
MR	Sepsis	25.69	21%
MR	Arthritis	24.78	33%
LR	Inflammation	46.98	47%
LR	Arthritis	27.7	40%
LR	Joint Diseases	27.22	41%
LR	Rheumatic Diseases	26.15	41%
LR	Gram-Negative Bacterial Infections	24.66	30%

Table 3.2.: Terms and Z-scores resulting from an ICA analysis and MeSH filtering. ER = early response; MR = middle response; LR = late response. Also the fraction of the genes associated with each MeSH term is given in %

largely governed by the pro-inflammatory cytokines (TNF, IL13, IL1B) and chemokines (CXCL2, CXCL3, CXCL5, CCL2-5, CCL8) as well as up-regulation of NFκB. This is followed by activation of TNFα and NFκB induced proteins like TRAF1, MMP9 and the major histocompatibility complex proteins HLA-DRB1, HLA-A and HLA-B. During late response, again the activity of the chemokines CXCL1 and CXCL5 were discovered, as well as the IL8 related genes MRC1, MX1 and CCL18. Here again, the accordance to the 54 top regulated genes is striking through a complete overlap of the associated highest ranked MeSH Terms: "Inflammation", "Arthritis", "Joint Diseases", "Bacterial Infections and Mycoses" and "Systemic Inflammatory Response Syndrome".

A further attribute of ICA based analysis is the grouping of genes into non-exclusive clusters. Hence, genes influencing more than one specific process can be found in more than one RG. Some of those interesting genes are the cytokines IL1B and IL8 or the surface protein coding genes CD36 and CD44 which were identified as presumably key players for gene regulatory networks involved in LVS infection response.

3.4. Discussion

Using the data-driven ICA approach, additional novel pathways were identified in addition to pathways similar to the ones deduced from classical hierarchical clustering approaches. Among the early responders, the pro-inflammatory cytokines TNFα and CCL2 were induced, which confirm previous findings about the secretion of large amounts of these inflammatory cytokines in a similar in vitro model using murine macrophages and human cell lines [Loegring et al., 2006]. Furthermore, in a murine macrophage cell line model, testing immediate responder genes by microarray analysis within the first 4 hours after infection with *F. tularensis* LVS, TNFα was found to be the main signal transducer whose expression level was found to be increased along with genes representing cytokine signaling-, enzyme- and transcription factor-families [Andersson et al., 2006]. The differences observed between our early responder genes and the immediate responders found in the murine model system emphasize the need of a multi-time point kinetic model of macrophage response to *F. tularensis* LVS infection with a well established microarray analysis method.

The virulence of *F. tularensis* depends on its ability to escape into the cytosol of the host cell, which reacts with the assembly of the caspase-1 dependent inflammosome complex. This process is closely related to the secretion of IL1b, IL18 and IL33, by which the induction of IL1b was also found with our analysis [Henry and Monack, 2007]. Recently, a natural killer (NK) cell cytokine, IFNγ dependent activation pathway was found to be relevant for the specific immune response to *F. tularensis* LVS infection [López et al., 2004]. We found a significant up-regulation of the IFNγ receptor 2 in macrophages, which in turn sensitizes these cells for the NK-cell derived IFNγ to result in a specific response.

These data show that, with the help of in vitro model systems using microarray analysis, the mechanism of *F. tularensis* LVS response can be well characterized and disease specific pathways discovered and identified. Moreover we could show that NFκB plays a major role regulating the immune response to *F. tularensis* LVS infection.

In comparison to the commonly used hierarchical clustering method, we found that our calculations using ICA resulted in higher clustering resolutions. The response specific MeSH terms derived through an ICA analysis are more closely related to the experiment (Bacterial Infections and Mycoses, Gram-Negative Bacterial Infections) and all three *response groups* show Inflammation as the most highly ranked MeSH term. Moreover, the nonexclusive clustering attribute of ICA leads to a more detailed insight into time-dependent patterns of the immune response.

4. Intronic microRNAs support their host genes by mediating synergistic and antagonistic regulatory effects

4.1. Introduction

Gene regulation via microRNAs (miRNAs), small ~22 nucleotide long RNA molecules, is a strongly conserved mechanism found in nearly all multicellular organisms including animals and plants [Carrington and Ambros, 2003]. Incorporated in a protein complex mainly built of Argonaute proteins, miRNAs bind preferably to complementary regions within the 3' UTRs of mRNAs, their target sites. About 37% of known mammalian miRNAs are located within the introns of protein coding genes, so-called host genes [Griffiths-Jones et al., 2006]. This has to be appreciated as a vague estimate since the amount of annotated miRNAs varies strongly from 117 for *bos taurus* to 695 for *homo sapiens*, and expectations of the functionally active fraction of the genome presumes amounts of miRNAs far above these numbers [Pheasant and Mattick, 2007; Birney et al., 2007]. For instance, the proportions for mouse (44%) and human (53%), two of the best studied mammals, were strikingly larger. Furthermore, intronic miRNAs appear to be conserved across several species [Ying and Lin, 2005; Rodriguez et al., 2004; Saini et al., 2008]. These miRNAs are transcriptionally linked to their host gene expressions and processed from the same primary transcript [Baskerville and Bartel, 2005]. Besides Drosha-processed miRNAs, a second type of intronic miRNAs, termed mirtrons, is known, that bypass Drosha cleavage by splicing [Ruby et al., 2007; Chan and Slack, 2007] but exhibit the same co-expression patterns with their host genes.

In animals, and more recently also in plants, it has been found that exact complementarity of target sites is not required for functional regulation. Unlike perfect matching,

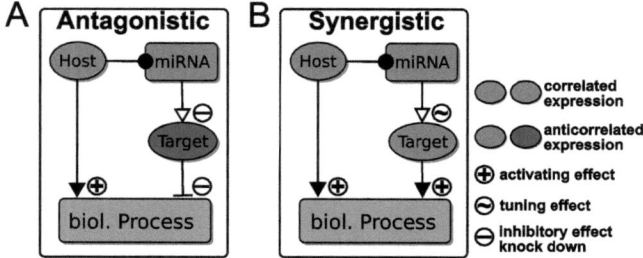

Figure 4.1.: The two proposed regulatory mechanisms of functional host to miRNA relationships. Genes are marked by ellipses, miRNA by rounded rectangles. Host miRNA relations are indicated by an edge with a dot. MicroRNA target regulation is indicated by a blank triangle. A minus denotes knock down of the target gene, whereas tilde denotes regulatory tuning. Activating effects on a biological process is shown by an arrowhead, inhibition is indicated with stops. Expression is color coded, co-expression is indicated green and anticorrelated expression red.(**A**) An antagonistic effect can be achieved by miRNA mediated downregulation of a gene with perturbing effect on a pathway or biological process regulated by the host gene. (**B**) Synergistic effect by miRNA mediated fine tuning of a target gene with common contribution of host and target gene to a pathway or biological process. The colored figure can be found at: http://www.helmholtz-muenchen.de/fileadmin/CMB/IMG/ImgDissLutter/Figure4-1.pdf

which leads to cleavage of the mRNA, partial complementarity of the target mRNA mainly leads to inhibition of ribosomal translation. However, due to the noncatalytic character of the miRNA-mediated regulation, both mechanisms have similar inhibitory effects [Levine et al., 2007]. MiRNA-mediated gene regulation can be categorized into 'switch', 'tuning' and 'neutral' [Bartel, 2004, 2009] effects. Switch regulation describes a knock-down of protein levels under a specific functional threshold caused by effective translational inhibition or cleavage of the target mRNA. In contrast, tuning does not inhibit target activity completely but tunes expression in a way such that miRNA targets are adjusted to a specific expression level required under specific cellular conditions. By neural targets one denotes miRNA-mRNA interactions, that are functional but without any advantageous nor adverse consequences to the cell. Since the neutral regulation does not have any effect on the phenotype, it will further on not be discussed in this work. MicroRNA-mediated regulatory mechanisms are known to appear in animals from early developmental stages to maturated adult tissues. They play a role in a variety of biological processes including cell differentiation, stem cell maintenance, proliferation as well as regulation of apoptosis [Stefani and Slack, 2008; Hwang and Mendell, 2006].

It is a common paradigm in biology that conservation on the genome level also implies a conservation of function. Therefore we hypothesize that the widespread appearance of the transcriptional junction of a protein coding gene and the regulatory miRNA implies a common function. Specifically, the co-regulation of a miRNA with its host gene may include two different main functions: (i) An antagonistic effect is achieved by miRNA mediated downregulation of genes with perturbing effects on a pathway or biological process activated by the host gene. The combined expression of an effector gene and a miRNA, which blocks translation of such antagonistic gene products, is a simple but elegant way to promote and support host gene functionality (Figure 4.1A). (ii) A synergistic effect is achieved by adjusting the protein expression levels of intronic miRNA targets towards intended optimal concentrations. A specific ratio between host and target gene products then allows for effective and optimized cooperative actions of co-regulated genes (Figure 4.1B). In humans, a functional relation between the host gene *GRID1* and the intronic miR-346 has been shown recently [Zhu et al., 2009] and the here proposed antagonistic effect has been proven for the intronic miR-338 and its host gene *AATK* [Barik, 2008].

In this work, we investigated the functional relation between miRNA host genes and putative targets of corresponding intronic miRNAs with a data-driven approach based on large-scale gene expression data and a knowledge-based approach using gene annotations. Genes sharing a common function, such as being involved in the same biological pathway, tend to share similar regulatory mechanisms and therefore appear as co-expressed genes in their expression profiles [Allocco et al., 2004]. Thus, genes with correlated time-dependent expression patterns are likely to be involved in functionally related cellular processes with synergistic effects. In contrast, anticorrelated expression pattern would promote the assumption that the participant genes take part in related, but antagonistic processes. Furthermore, functional gene annotations as provided by the Gene Ontology (GO) [Ashburner et al., 2000] give information about a common or strongly related function of two genes, for instance hosts and targets. We hypothesized that functional relations between miRNA host genes and related target genes appear in significant correlated expression patterns and we expected, that host and target gene sets are closer related in the GC as randomly sampled sets, for both antagonistic and synergistic motifs as introduced in figure 4.1.

4.2. Results and Discussion

4.2.1. Targets of similarly expressed host genes show correlated expression patterns

We studied the relationship between host and target genes, in three different mouse developmental microarray datasets (see methods): embryonic stem cell development (SCD), somitogenesis (SG) and neurite outgrowth (NO). We chose developmental datasets since regulatory effects of miRNAs are known to be strongly present in developmental processes [Gangaraju and Lin, 2009]. During cell differentiation, groups of genes driving specific developmental processes are often commonly regulated, arising in the phenotypic effect of similar expression patterns of these genes in time course data. A synergistic relationship between host and the miRNA target genes of differentiating cells is then indicated by positively correlated gene expression patterns. In reverse, antagonistic processes are expected to show anticorrelated or weakly correlated expression patterns between host and related target genes.

Since we argue that correlated expression indicates for potential common host gene functions, we initially tested for correlations between host gene expressions. In order to generate statistically robust results, independent of data and prediction errors, we did not analyze single gene expression patterns but argue on groups of correlated genes. Therefore, for each dataset we identified all miRNA host genes and clustered their time courses according to correlations above 0.8 (see methods). Within all analyzed cell differentiation datasets, host genes tend to be co-expressed in clusters. As a result of our clustering we obtained seven host gene clusters with more than 5 host genes (see table 4.1).

Intriguingly, some host genes appear to be clustered together preferentially across the experiments. The genes *H19, Igf2, Lpp, Plod3*, and *Rnf130* were clustered together in the two clusters SCD I and NO I, and the genes *Chm, Copz1, Dnm1, Nupl1*, and *Sf3a3* together in the clusters SG I and NO II.

For each host gene cluster we identified the intronic miRNAs and all their expressed targets. Most prediction tools for miRNA target site prediction vary in qualitative and quantitative manner. In order to get more confident predictions, we used a consensus model (C) of several miRNA target prediction tools (see methods). A detailed list of all analyzed miRNAs/clusters in this work including host genes, loci, a correlation and a

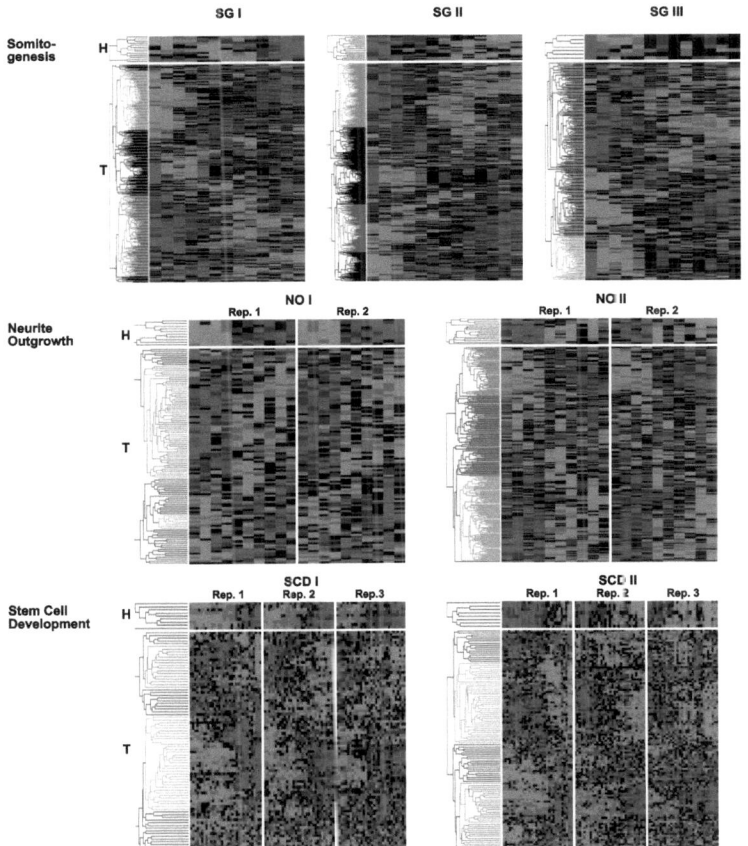

Figure 4.2.: Clustered heat maps for the seven host gene cluster (**H**) and the corresponding target gene expression profiles (**T**). For all three time course datasets only clusters with more than five host genes are shown. Each row corresponds to one gene expression pattern, each column to a measurement. Time dependent measurements are shown in ascending order from left to right. The expression level of each gene is standardized so that the mean is set to 0 and the standard deviation is 1. Expression levels above and below 0 are color-coded; red indicated for high and green for low expression levels, respectively; black for zero expression values. Biological replicates of the three datasets are in order from Rep. 1 to Rep. 2 and Rep. 3, respectively. Hierarchical clustering with euclidean distance metric and average linkage is used. Colored subtrees in the dendrogramm denote for co-expressed (green) or anticorrelated (red) gene expression of predicted targets. (**Somitogenesis**) The dataset splits up into three host gene cluster, SG I with 13, SG II with 21, and SG III with 7 host genes. (**Neurite Outgrowth**) Two cluster with 10 (NO I) and 17 (NO II) host genes could be identified with similar behaviour of host and target genes in both replicates. (**Stem Cell Development**) Two host gene clusters containing 9 (SCD I) and 8 (SCD II) hostgenes were identified. All host and target genes show similar behaviour in all three replicates. For each dataset, flipped expression patterns between the host/target clusters are striking (SG I vs. SG II; NO I vs. NO II; SCD I vs. SCD III). The colored figure can be found at: http://www.helmholtz-muenchen.de/fileadmin/CMB/IMG/ImgDissLutter/Figure4-2.pdf

GO similarity based score is available as Supplementary Table 1.

For the seven clusters we performed a hierarchical cluster analysis based on the expression data of the target genes (see Figure 4.2). All resulting trees mainly split up in two subclusters: one subcluster of genes with similar or positively correlated expression patterns and one with opposing or anticorrelated expression compared to the host genes, respectively. Furthermore, within each dataset, the resulting trees of at least two target gene groups appeared to show completely flipped expression patterns of the main subclusters (SG I vs. SG II; NO I vs. NO II; SCD I vs. SCD III).

These results fit well to the observation that miRNAs dampen the output of preexisting mRNAs or optimize required protein output as it is proposed for metazoans [Bartel and Chen, 2004]. Additionally, in [Farh et al., 2005] it was shown that genes preferentially expressed at the same time and place as a miRNA tend to avoid sites matching the miRNA. By contrast, co-expression of a transcripts with evolutionary conserved miRNA binding site would then arise from a functional requirement.

The clear discrimination between the two expression patterns suggests a gradual order of differentiating cells, whereas miRNAs function as enhancers of robustness in gene regulation [Rhoades et al., 2002; Tsang et al., 2007]. A plausible explanation would be that shortly after initiation of the differentiation process, genes that arrange the differentiating cell towards its new function are up-regulated. In this stage miRNAs are activated to inhibit processes required for self-renewal of stem cells but were perturbed during differentiation. After this 'reprogramming' the cell adopts new functions and stabilizes. In this phase genes are up-regulated which now fulfill the cell's new responsibilities and simultaneously block activity that was only required for differentiation.

4.2.2. MicroRNA host gene cluster and related target genes show significant correlations of their expression patterns and functional similarities

In order to confirm the above observed and to show statistically that gene expression patterns of host genes are significantly correlated with the patterns of their predicted target genes, we determined the correlation distribution for each cluster by calculating correlation coefficients between all hosts and all expressed putative target genes. These distributions were compared to 500 sets of randomly sampled target genes (see methods).

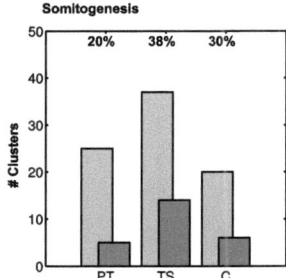 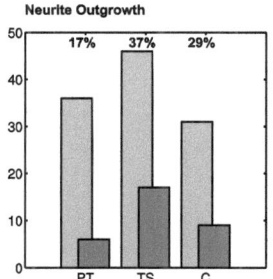 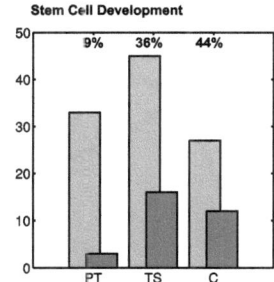

Figure 4.3.: Results of the host gene cluster based expression analysis. Grey bars denote the number of all identified host gene clusters including unclustered hosts with expressed target genes, predicted by Pictar (PT), TargetScan (TS) and our consensus model (C). Orange bars denote the number of clusters with significantly correlated target gene expression patterns. The relative fraction of significant clusters for each dataset and miRNA target prediction tool is denoted. The colored figure can be found at: http://www.helmholtz-muenchen.de/fileadmin/CMB/IMG/ImgDissLutter/Figure4-3.pdf

To avoid any bias by our consensus model, we additionally used two further independent prediction tools, namely Pictar (PT) [Krek et al., 2005] and TargetScan (TS) [Lewis et al., 2003]. For each host gene cluster and each single host gene, expression patterns were compared to expression of predicted targets. Only clusters with predicted and expressed targets in the respective dataset were used in the following analyses.

Results can be seen in Figure 4.3. Concordant for all used methods and all analyzed datasets, we determined that up to 44% of the identified host gene clusters were significantly positively correlated or anti-correlated to their target gene expressions. Comparing the the three datasets, we only found marginal differences. The average amount of host gene clusters with significant correlated target expression varies between 27% and 30%.

Comparing the three tools, PT performs strikingly weaker (15%) than TS and the consensus model with regard to the mean fraction of host gene clusters with significant correlated predicted target expressions (37% and 34%). Since the number of targets predicted with PT for each host gene is in average considerably smaller compared to the two other methods, false positive predictions have a larger effect on the determined p-values.

Taking into account that the consensus model graph is less dense as well as noteably smaller than the TS graph, it performs best in this analysis with an equal fraction of

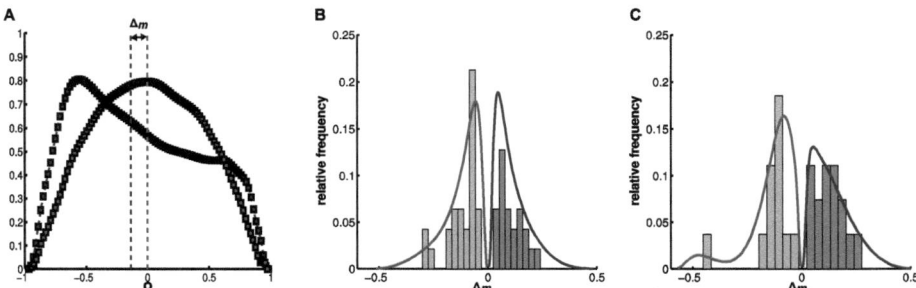

Figure 4.4.: As an example, figure(**A**) shows the distributions of correlation coefficients ρ between host and target gene expression patterns (**blue**) of Cluster NO I and correlation coefficients ρ between the same host genes and sampled target genes (**red**). The medians are illustrated by blue and red lines, respectively. Δ_m indicates the difference between the two medians. A missing relation between host and target gene expression would result in a difference $\Delta_m = 0$. The distributions of Δ_m taken over all significant clusters of the three datasets are shown in the two histograms for targetscan (**B**) and our consensus model (**C**). Estimated densities of positive and negative Δ_m distributions indicating for antagonistic or synergistic regulatory effects are shown by the orange and green line. Missing distances of $\Delta_m = 0$ in both distributions indicate that all significant clusters deviate from the null model (sampled data). Both distributions show bimodal shape with equal maxima on both sides, indicating that positive and negative correlations are approximately equally distributed over all analyzed clusters. The colored figure can be found at: http://www.helmholtz-muenchen.de/fileadmin/CMB/IMG/ImgDissLutter/Figure4-4.pdf

significantly regulated clusters. However, our results are consistent over all datasets and all different miRNA target site prediction tools.

4.2.3. Functional relation between host and target genes includes synergistic as well as antagonistic effects

The previously shown results so far indicate a nondirectional functional relation between host genes and intronic miRNAs, but do not provide any information on positive or negative correlations. Since these results show that PT predictions agree with the two other tools, but due to the small size of the graph and therefore its lack of robustness, we excluded PT from this analysis.

To test whether one or both of the two proposed functional effects — synergistic or antagonistic — may be identified in our data, we calculated the distance between the medians of the correlations between the host and predicted target genes and the correlations between the hosts and randomly sampled targets (Figure 4.4A and methods). The resulting distances Δ_m combined from all three datasets can be seen in Figure 4.4B and 4.4C. Both distance distributions show a bimodal distribution with a local minimum at $\Delta_m = 0$, but no significant shift towards a negative or positive correlation. Hence, based on the assumption that high positive or negative correlation of gene expression patterns indicated similar or opposite functions, we infer that both proposed effects, knock down and fine tuning, appear to be equally represented in our data.

Since our investigation is only based on mRNA expression data and further information on protein levels is missing, the real impact on translation stays obscure in this analysis. However, in [Baek et al., 2008] it could be shown that most of miRNA-mRNA interactions function as fine scaling adjustments to the proteome. Considering the fact that our experimental analysis was only based on mRNA expression data, only knock down effects are directly visible. But in agreement with previous work [Selbach et al., 2008], the massive appearance of positively correlated miRNA and target expression strongly indicates tuning effects of varying translational repression.

4.2.4. Host and target gene sets display enriched functional similarity

The significantly correlated expression patterns between host genes and miRNA target genes support the notion that intronic miRNA regulation improves host-associated biological functions by either tuning or dampening the expression of target genes. We assume that this relation is also apparent via shared functional annotations. To test this hypothesis, we determined the commonly used functional similarity of gene products based on Gene Ontology (GO) [Schlicker et al., 2006] between a single or multiple host genes and their set of target genes. We then calculated the significance of the mean functional similarity by comparing the target set with randomly sampled sets of miRNA target genes (see methods).

We analyzed the previously defined clusters SCD I – NO II and calculated mean functional similarities between the host and target gene sets. Results are shown in Table 4.1. All host gene clusters display a significantly higher functional similarity ($p < 0.05$) to their predicted TS target genes as compared to the null model of randomly chosen target genes. Only the two clusters SCD I and SCD II exceed the significance level of

	Hosts	PT targets	p	TS targets	p	C targets	p
SCD I	9	7	0.0305	275	$< 10^{-4}$	82	0.1425
SCD II	8	68	0.0578	771	$< 10^{-4}$	109	$< 10^{-4}$
SG I	13	149	$< 10^{-4}$	1521	$< 10^{-4}$	377	$< 10^{-4}$
SG II	21	189	$< 10^{-4}$	1956	$< 10^{-4}$	486	$< 10^{-4}$
SG III	7	39	$< 10^{-4}$	617	0.0008	258	$< 10^{-4}$
NO I	10	51	0.0016	864	0.0109	112	$< 10^{-4}$
NO II	17	67	0.0046	1274	$< 10^{-4}$	218	$< 10^{-4}$

Table 4.1.: Host gene cluster size and number of target genes, predicted with the three methods Pictar (PT), TargetScan (TS), and our Consensus model (C), respectively. The p-values determined by a comparison of functional GO similarities between host and predicted targets to randomly chosen sets of target genes of identical size are shown.

0.05 for consensus model and PicTar predictions, respectively.

To check whether a high functional similarity can be found for all host-target relations independent of expression patterns, we additionally calculated the functional similarity score for all host genes and their predicted target gene sets. We expected the most robust results for the largest network of predicted microRNA target gene associations, since the score is given by the mean of all host gene - target gene pairs. In Figure 4.5A, we plotted the frequency distribution of similarity scores for TS. We found that the scores are well distributed within the range of 0 and 5. We compared each similarity score with a null model, where the same number of target genes is randomly selected from all miRNA target genes as provided by TS. For the host gene *Copz1* for example, we found a significantly larger functional similarity to its targets as compared to 1000 randomly selected sets of microRNA targets (see Figure 4.5B).

For all annotated host genes with available annotations for the respective targets, we calculated p-values and z scores, as measures of deviation from the null model. We found that surprisingly many host-target relations deviated from the null model, with high z scores as can be seen in Figure 4.5C. As many as 57 of all 75 host genes annotated in the ontology 'biological process' exhibited a greater similarity to their targets ($z > 0$) than expected by chance, 30 of them with a p-value < 0.05. For those pairs of host and target genes, a strong correlation in terms of their annotated 'biological process' existed. For the other prediction tools used on in this study, a similar trend to high z scores could be observed (see supplementary figure 1). However, these predictions comprise

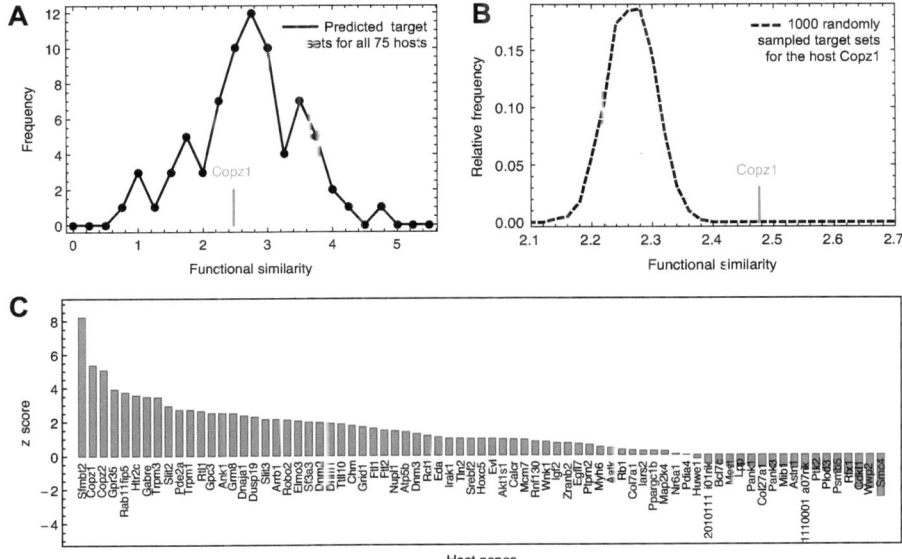

Figure 4.5.: Functional similarity of host and target gene sets as predicted by TargetScan. (**A**) Frequency distribution of the functional similarity score for all 75 host-target relations. For each single host gene and its set of target genes, we calculate a mean score based on the GO annotation 'biological process'. The mean functional similarity of the host gene Copz1 to its predicted targets is 2.48 (blue line). (**B**) Comparison of the real functional similarity score the host gene Copz1 with a null model distribution. For the null model, a random set of microRNA target genes of the same size has been chosen 1000 times and the functional similarity score has been calculated. The real score of Copz1 deviates significantly from the null model distribution, resulting in a high z score. (**C**) Z scores for all annotated host genes. A total of 21 out of 75 host genes show z scores > 2 and thus display a significantly higher functional similarity as expected from a random sample of target genes.

less annotated host genes (48 and 45 for PicTar and consensus model, respectively) and also about 10 times less links, rendering significant deviations less possible (see methods for details).

With the use of GO gene annotations we could show that intronic miRNA tend to target genes that are functionally more similar to the host genes than randomly chosen genes. The strong bias towards positive correlations and absence of significant dissimilarities agrees with both former proposed regulatory principles (figure 4.1A,B).

Notably, GO terms are not classified on their antagonistic effects on each other but on biological relations. For instance, two pathways with conflicting regulation on a cellular process like 'cell growth' are both children of the parental term and therefore close within the GO tree. Furthermore, two genes can have opposed regulatory effects on one pathway and would be still grouped together in the same term.

4.3. Conclusion

The results of this work show that the genomic linkage between intronic miRNAs and their host genes coincides with a functional relation. Using a data-driven as well as a knowledge-based approach, miRNA host genes and related target genes were analyzed towards functional relations. Expression patterns were obtained from three developmental datasets. Correlated expressions of host and miRNA target genes deviated significantly from a random model. Both, positively and negatively correlation patterns have been observed in approximately equal amounts. An independent GO analysis of the predicted miRNA-mRNA interaction network confirmed that host and predicted target genes tend be annotated with similar or related terms, compared to a random model. Taken together, results indicate for either synergistic or antagonistic regulatory effects mediated by either downregulation of genes with an opposed function or fine-tuning of miRNA targets, co-operative to the host gene.

4.4. Material and Methods

4.4.1. Microarray data and preprocessing

All analyzed datasets were taken from the GEO [Barrett and Edgar, 2006] database: (i) The stem cell development (SCD) datasets consists of three cell lines (R1, J1, V6.5) differentiated into embryoid bodies (EB) at 11 time points from t=0h until t=d14. From each time point and each cell line 3 technical replicates were measured (combination of three cell line differentiations GSE2972, GSE3749, GSE3231). (ii) Within the somitogenesis dataset (SG) gene expression was measured from synchronized C2C12 myoblasts at 13 timepoints from t=0h until t=6h (GSE7012). (iii) The neurite outgrowth (NO) and regeneration dataset consists of transcriptional activity, measured from dorsal root ganglia during a time course of neurite outgrowth in vitro under two conditions: untreated and under potent inhibitory cue Semaphorin3A. Measurements were taken at 5 time points from t=2h until t=40h including two technical replicates (GSE9738).

Affymetrix raw data were preprocessed using Bioconductor's R package *simpleaffy* [Wilson and Miller, 2005]. Data was normalized and detection calls were determined. Expression values were calculated using the RMA algorithm. Each dataset was filtered independently to remove all probesets with an absent flag in more then two third of all datapoints within the whole experimental setup.

Gene names and gene symbols for each probeset were derived from the Bioconductor Affymetrix Mouse Expression Set 430 annotation data (moe430a.db). Gene symbols represented by more than one probeset were set to the median expression values.

4.4.2. Expression profile based analysis

Host gene cluster were defined upon a correlation-based adjacency matrix. For each microarray dataset we selected all known miRNA host genes and calculated a correlation matrix based on their expression profiles. Each entry representing a correlation coefficient above 0.8, was set to 1, all others to 0. This adjacency matrix now forms a graph of host genes. A host gene cluster was then defined as a maximal connected subgraph of this graph. This equals nearest neighbour method applied to hierarchical clustering algorithm with a defined cutoff of 0.8 of the dendrogramm. For each host gene cluster containing M host genes, the N corresponding target genes were determined upon the three miRNA target prediction tools.

We calculated the cluster specific miRNA degree $d_i = \#T_i/\#H_i$ where $\#T_i$ is the number of target genes and $\#H_i$ the number of host genes of cluster i.

Depending on the respective expression profiles, we calculated the $M \times N$ cross-correlation coefficients between all hosts and all targets. As a null model we randomly sampled N targets for 500 times. For each sample we calculated all $M \times N$ correlations. Statistically significant differences between the correlation distributions of our clusters and sampled data were estimated by determining p-values using Wilcoxon's rank sum test.

Distances between the medians of the correlation distributions were calculated as

$$\Delta_n = median(C_c) - median(C_s) \qquad (4.1)$$

with C_c being the correlation distribution between the host and the target genes of one cluster and C_s being the correlation distribution between host genes and sampled target genes of one cluster.

Hierarchical cluster analysis was performed using Matlab's Bioinformatics toolbox (http://www.mathworks.com) using average linkage with Euclidean distance metric.

4.4.3. Intronic miRNAs and target prediction

A list of all murine intronic miRNAs and their host genes was downloaded from the miRBase website (http://microrna.sanger.ac.uk). Predictions made by PT were downloaded from the UCSC genome browser (http://genome.ucsc.edu) and TS conserved miRNA target site predictions were downloaded from the TS website (http://www.targetscan.org). Redundant gene to miRNA relationships were removed from both datasets.

The Consensus model prediction graph used in our analysis was built of five different miRNA target site prediction tools. Additionally to PT and TS we used predictions from PITA [Kertesz et al., 2007], Miranda [Betel et al., 2008] (http://www.microrna.org), and targetspy (Sturm et al, submitted). From all predictions based on RefSeq transcript IDs, we filtered out only miRNA-transcript relations that were predicted by a minimum of four different tools. Transcript mapping to gene symbols was done using a local copy of the RefSeq database (September 2008) [Pruitt et al., 2007].

These genome-wide predictions can be represented by a network (bipartite graph), where the two different sets of nodes are formed by the miRNAs and the target genes, respectively, and the predicted interactions are formed by the edges. The three graphs vary primarily in their absolute sizes. PT with 242 miRNAs and 1335 overall predicted targets is very small compared to TS (382 miRNAs, 8879 targets) and consensus model (219 miRNAs, 3249 targets). In figure 4.6A relative densities for all graphs and in figure 4.6B all degree distributions are shown. For each cluster a mean miRNA target recovery was calculated as the fraction of the number of all predicted and recovered target genes of one cluster through the number of clustered host genes. These distributions again are strikingly similar whereas the mean still varies strongly (Figure 4.6C,D).

The fraction of the cluster-specific miRNA degree compared to the complete graph miRNA degree of consensus model is very high (76%) compared to the other methods (TS: 50%, PT: 27%). Since TS predicts the highest number of targets per miRNA, one also expects a relatively huge recovery of target genes within the dataset. The PT

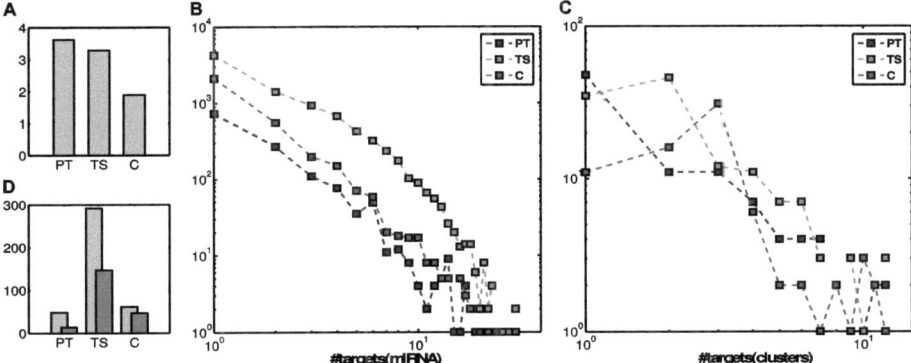

Figure 4.6.: Properties of the three miRNA-target bipartite graphs. (**A**) The relative densities, number of existing edges divided by all possible edges, in percent of the three graphs for Pictar (PT), TargetScan (TS), and consensus model (C). (**B**) Log-log plot of number of predicted miRNA targets for all three different prediction graphs. (**C**) Log-log plot of cluster specific miRNA target recovery for all three different prediction graphs (for details see text). (**D**) The mean of the numbers of predicted miRNA targets of the complete graphs (grey), and cluster-specific recovery of miRNA targets (orange): Mean of the sums of all identified targets of one host gene cluster divided by the sums of all host genes of the cluster. The colored figure can be found at: http://www.helmholtz-muenchen.de/fileadmin/CMB/IMG/ImgDissLutter/Figure4-6.pdf

graph is the densest graph of all but also the smallest one, hence the weak recovery of targets. One reason for the high target recovery of the consensus model might be that the used prediction tools for the consensus model score are all trained upon validated data. Therefore, the resulting miRNA-target predictions contain more training data as the PT and TS, which results in the high recovery rate.

4.4.4. Functional similarity of host genes and target gene sets

We assume that host genes confer regulatory control by translational inhibition of the respective intronic microRNA target genes in possibly related biological processes. To test this hypothesis for all hosts and target genes, we compare the similarity of their respective annotations. Functional gene annotations as provided by the GO [Ashburner et al., 2000] classify genes according to their function, associated biological processes or appearance within defined cellular components. They are organized hierarchically, typically in a directed acyclic graph. To each gene more than one classification term can be assigned.

The functional similarity between a host and a target was defined by *Resnik's measure* as described in [Schlicker et al., 2006] and calculated using the ProCope software suite [Krumsiek et al., 2008]. This method scores relationships between genes by common appearance within one or more terms or, more abstract, by analyzing their distance within the GO graph. For genes with multiple term annotations the maximum scoring GO term pair was used. The functional similarity between a host and a set of targets was determined as the mean of all single host-target scores. For our study, we downloaded the most recent GO files and mouse gene annotation lists from the GO website (January, 2009).

In order to assign statistical significance of the functional host-target similarities in our network, we compared the average similarity of each host to all of its targets against 10.000 randomized networks. To evaluate the host-cluster to target relations we compared the average host-target similarities in the real network against 10.000 networks with randomized target sets for each host cluster. We calculated a p-value as the relative number of samples with higher scores. The z score was calculated as the deviation of the real score s from the mean m of the sampled distribution, divided by its standard distribution σ, $z = \frac{s-m}{\sigma}$.

5. Discussion

The analysis of gene expression is a challenging task in modern biology. Microarray technology allows for large-scale measurements of the expression of thousands of genes at once. Based on different biological models, various methods have been developed to analyze these data in an appropriate manner. In this work we distinguished between mapping and mixing models. Mapping models compare the expression values of different measurements directly in order to either identify differentially regulated genes, prevalently appearing expression patterns or to extract potential marker genes. In contrast, mixing models follow a different approach. The underlying assumption here is that a single gene expression profile is composed of several superimposing expression modes. These expression modes therefore represent specific biological processes responsible for a distinct cellular task.

The different methods applied range from classical statistic approaches, such as the t-test, clustering methods like hierarchical clustering, to methods developed in linear algebra like ICA. In this work two of these methods, ICA and hierarchical clustering were applied to different microarray datasets following diverse biological questions.

In chapter 2 two different classes of gene expression profiles, derived from monocytes and M-CSF dependent differentiated macrophages, were analyzed. Statistically independent GEM were extracted from the observed expression profiles using ICA. From each GEM a group of genes was deduced, henceforth called *sub-mode*. These *sub-modes* were further analyzed with different database query and literature mining tools and then combined to form so called *meta-modes*. With these a knowledge-based pathway analysis was performed and a well-known signal cascade could be reconstructed. Although there exists lot of other work applying ICA to microarray data [Liebermeister, 2002; Lee and Batzoglou, 2003; Chiappetta et al., 2004], a detailed biological discussion of the results is mostly missing. In this work, a special focus was to test the ICA derived results for biological relevance, and to provide a reasonable approach for interpretation.

The results show that ICA is an appropriate tool to uncover underlying biological mechanisms from microarray data. Most of the well known pathways of M-CSF dependent monocyte to macrophage differentiation could be identified by this unsupervised microarray data analysis. Moreover, recent research results like the involvement of proliferation associated cellular mechanisms during macrophage differentiation, could also be corroborated.

Chapter 3, again deals with the application of ICA to microarray data. However, in contrast to chapter 2 ICA was applied to kinetic gene expression profiles and compared to the more commonly used method of hierarchical clustering. The dataset consisted of human monocyte derived macrophages from three different donors infected with the intracellular pathogen *Francisella tularensis*.

Results were compared using pathway analysis tools, based on the Gene Ontology and the MeSH database. It could be shown that both methods lead to time-dependent gene regulatory patterns, which fit well to known TNFα induced immune responses. In comparison, the nonexclusive attribute of ICA results in a more detailed view and a higher resolution in time dependent behavior of the immune response genes. Additionally, NFκB could be identified as one of the main regulatory genes during the response to *F. tularensis* infection.

A less methods and more biology oriented approach applied to microarray data was discussed in chapter 4. Since up to 53% of mammalian miRNAs appear to be located within introns of protein coding genes, the linkage between their expression and the promoter-driven regulation of their host gene was analyzed. Therefore, the study investigated this linkage towards a relationship beyond transcriptional co-regulation. Using measures based on both annotation and experimental data, it could be shown that host genes and their intronic miRNAs are often functionally related.

The study showed that miRNA target genes tend to show expression patterns significantly correlated with the expression of their host genes. By calculating functional similarities between host and predicted miRNA target genes based on GO annotations, it could be confirmed that many miRNAs link the annotated function of their host genes with that of the target genes. Additionally, these results indicate that miRNAs support host gene activity in an either synergistic or antagonistic manner.

The result of the latter analysis also brings a new perception to the analysis of gene expression data. The so far common paradigm to map these large scale data to protein associated gene functionality may be extended by the knowledge of intronic miRNA functionality. This is especially interesting since most commonly used microarrays are not only limited to the detection of protein coding genes, but can also deliver information about further non-coding gene regulatory mechanisms.

Taken together, this work shows that the analysis of microarray data, depending on the applied method can lead to diverse biological findings. Thereby, it is less important to use the most sophisticated tool, but more important to carefully reflect on the aim of the experiment to choose the appropriate method. However, developing new analysis tools or assigning methods developed in other fields to the analysis of large scale gene expression data is still a demanding task. On the one hand, it is still necessary to overcome open problems like noise reduction in the data and, on the other hand, to improve the biological outcomes and to provide more meaningful results.

Furthermore, in some cases a specific tool outperforms a commonly used one as shown in chapter 3, and should therefore be preferentially used. Hence, newly developed methods have to be proven carefully in their applicability and compared to commonly used ones.

Finally, we now will give a short outlook on how to overcome several issues, improve diverse methods and on future directions in large scale gene expression data analysis. For instance, investigations on ICA algorithms, including subspace analysis, will allow for a more adapted mixing model of the underlying biological processes [Gruber et al., 2009]. Remaining dependencies of extracted biological processes may be identified, hence delivering a more adapted view on large regulatory networks. Further exploration of the mixing coefficients derived with matrix factorization methods, as described in chapter 3 or in [Schachtner et al., 2008], from time dependent data may carry out improvements in the reconstruction of time dependent regulatory networks. A promising idea is the application of non-negative tensor factorization (NTF) methods [Cichocki et al., 2007], that allow for the use of higher dimensionality in the mixing model. Thus, varieties within biological replicates may be identified. For instance, processes running in different cell lines with different rates may be more precisely reconstructed.

As the knowledge about miRNAs and the comprehension of their impact on gene expression grows, the perspective on analyzing mRNA data will certainly change. Therefore, existing theoretical analysis methods have to be extended and development of new tools should be conducted with respect to this entity. As basically presented in this work, the use of graph theoretic approaches and network representation exhibit a promising approach to future investigations. Using information from diverse sources, such as microarray analyses, TF and miRNA regulatory networks, as well as gene annotations, large regulatory networks can be created. These networks can then be further analyzed and optimized, as for instance by predicting missing links via applying a Boolean approach. Following a modern systems biology approach with a high crosstalk between experimental and theoretical work, will also lead to improvements concerning the biological relevance of these models. Small subnetworks and motifs will give rise to pointed experiments that may in turn be used to upgrade the models.

Finally, recently developed experimental methods, like deep sequencing, will certainly improve the quality of gene expression data. This relatively new open method allows for large-scale measurements of the transcriptome, independent of the RNA type and may also deliver information about so far unknown transcripts. However, this technique will also demand new model assumptions and sophisticated analysis techniques.

In conclusion, we want to point out that the wide field of transcriptome analysis still offers lots of starting points for new investigatory approaches leading to further findings that will extend our understanding of the gene regulatory machinery forming a complete organism out of a single fertilized egg cell.

A. Monocyte/macrophage differentiation meta-modes

Table 3 - Signal transduction genes

SYMBOL	NAME	Pathway	CLU	ProbeSet
ADAM10	ADAM metallopeptidase domain 10	Cell C.	6.2	214895_S_AT
ADM	adrenomedullin	MAPK	12.2	202912_AT
AIF1	allograft inflammatory factor 1		13.2	209901_X_AT
ALDH1A1	aldehyde dehydrogenase 1 family, member A1		12.2	212224_AT
ARF1	ADP-ribosylation factor 1	Cell C.	6.2	208750_S_AT
ARFGEF1	ADP-ribosylation factor guanine nucleotide-exchange factor 1(brefeldin A-inhibited)		6.2	216266_S_AT
ATF3	activating transcription factor 3	MAPK	12.2	202672_S_AT
BCL3	B-cell CLL/lymphoma 3		3.2	204908_S_AT
BID	BH3 interacting domain death agonist		3.2	204493_AT
BIRC1	baculoviral IAP repeat-containing 1		13.2	204860_S_AT
BLNK	B-cell linker		13.2	207655_S_AT
C1QR1	complement component 1, q subcomponent, receptor 1		13.2	202878_S_AT
CAMKK2	calcium/calmodulin-dependent protein kinase kinase 2, beta		3.2	212252_AT
CASP1	caspase 1, apoptosis-related cysteine peptidase (interleukin 1, beta, convertase)	MAPK	12.2	211368_S_AT
CCL3	chemokine (C-C motif) ligand 3	MAPK	12.2	205114_S_AT
CD163	CD163 antigen		13.2	215049_X_AT
CD36	CD36 antigen (collagen type I receptor, thrombospondin receptor)		13.2	206488_S_AT
CD44	CD44 antigen (homing function and Indian blood group system)	MAPK	3.2	217523_AT
CD58	CD58 antigen, (lymphocyte function-associated antigen 3)		6.2	216942_AT
CD83	CD83 antigen (activated B lymphocytes, immunoglobulin superfamily)	MAPK	12.2	204440_AT
CD86	CD86 antigen (CD28 antigen ligand 2, B7-2 antigen)		13.2	205686_S_AT
CFLAR	CASP8 and FADD-like apoptosis regulator	Cell C.	6.2	211317_S_AT
CSPG2	chondroitin sulfate proteoglycan 2 (versican)		12.2	221731_X_AT
CTSK	cathepsin K (pycnodysostosis)	MAPK	12.2,13.2	202450_S_AT
CXCL1	chemokine (C-X-C motif) ligand 1 (melanoma growth stimulating activity, alpha)	MAPK	12.2	209774_X_AT
CYP1B1	cytochrome P450, family 1, subfamily B, polypeptide 1		3.2,6.2	202435_S_AT
DDX3X	DEAD (Asp-Glu-Ala-Asp) box polypeptide 3, X-linked		6.2	212514_X_AT
DUSP1	dual specificity phosphatase 1	MAPK	12.2	201041_S_AT
EGR2	early growth response 2 (Krox-20 homolog, Drosophila)	MAPK	3.2	205249_AT
EREG	epiregulin		3.2	205767_AT
FABP5	fatty acid binding protein 5 (psoriasis-associated)		13.2	202345_S_AT
FCGR1A	Fc fragment of IgG, high affinity Ia, receptor (CD64)		13.2	214511_X_AT
FLI1	Friend leukemia virus integration 1		13.2	204236_AT
G6PD	glucose-6-phosphate dehydrogenase	MAPK	3.2	202275_AT
GADD45B	growth arrest and DNA-damage-inducible, beta	MAPK	12.2	209305_S_AT
GDI2	GDP dissociation inhibitor 2	Cell C.	6.2	200008_S_AT
H2BFS	H2B histone family, member S		3.2	209806_AT
HOMER3	homer homolog 3 (Drosophila)	Cell C.	6.2	215489_X_AT
HPSE	heparanase		13.2	219403_S_AT
IER3	immediate early response 3		12.2	201315_X_AT
IFITM3	interferon induced transmembrane protein 3 (1-8U)		12.2	201315_X_AT
IGFBP7	insulin-like growth factor binding protein 7	MAPK	12.2	201163_S_AT
IL1RN	interleukin 1 receptor antagonist	MAPK	3.2	212657_S_AT
IL2RG	interleukin 2 receptor, gamma (severe combined immunodeficiency)		3.2	204116_AT
IL8	interleukin 8	MAPK	3.2,12.2	202859_X_AT
JUNB	jun B proto-oncogene	MAPK	12.2	201473_AT
KLF10	Kruppel-like factor 10		3.2,12.2	202393_S_AT
LMO2	LIM domain only 2 (rhombotin-like 1)		13.2	204249_S_AT
M6PR	mannose-6-phosphate receptor (cation dependent)		6.2	200900_S_AT
MAP3K2	mitogen-activated protein kinase kinase kinase 2	Cell C.	6.2	221695_S_AT
MAPK1	mitogen-activated protein kinase 1	MAPK	3.2	208351_S_AT
MCP	membrane cofactor protein (CD46, trophoblast-lymphocyte cross-reactive antigen)		6.2	207549_X_AT
MGLL	monoglyceride lipase		12.2	211026_S_AT
MMP9	matrix metallopeptidase 9 (gelatinase B, 92kDa gelatinase, 92kDa type IV collagenase)		13.2	203936_S_AT
MTSS1	metastasis suppressor 1		13.2	203037_S_AT
MYCL1	v-myc myelocytomatosis viral oncogene homolog 1, lung carcinoma derived (avian)		13.2	214058_AT

continued on next page

continued from previous page

SYMBOL	NAME	Pathway	CLU	ProbeSet
NEK3	NIMA (never in mitosis gene a)-related kinase 3		13.2	211089_S_AT
NFKBIE	nuclear factor of kappa light polypeptide gene enhancer in B-cells inhibitor, epsilon		3.2	203927_AT
OGT	O-linked N-acetylglucosamine (GlcNAc) transferase	Cell C.	6.2	207564_X_AT
PDE4B	phosphodiesterase 4B, cAMP-specific (phosphodiesterase E4 dunce homolog, Drosophila)		12.2	203708_AT
PECAM1	platelet/endothelial cell adhesion molecule (CD31 antigen)		13.2	208981_AT
PLEK	pleckstrin		12.2	203471_S_AT
PPP1R15A	protein phosphatase 1, regulatory (inhibitor) subunit 15A		12.2	37028_AT
PRNP	prion protein (p27-30)		6.2	215707_S_AT
PSEN1	presenilin 1 (Alzheimer disease 3)	Cell C.	6.2	207782_S_AT
PTGER2	prostaglandin E receptor 2 (subtype EP2), 53kDa		12.2	206631_AT
PTPRO	protein tyrosine phosphatase, receptor type, O		13.2	208121_S_AT
RALGDS	ral guanine nucleotide dissociation stimulator	MAPK	12.2	209050_S_AT
RIPK2	receptor-interacting serine-threonine kinase 2	MAPK	12.2	209545_S_AT
RPS6KA1	ribosomal protein S6 kinase, 90kDa, polypeptide 1	MAPK	3.2	203379_AT
S100A8	S100 calcium binding protein A8 (calgranulin A)	MAPK	12.2	202917_S_AT
S100A9	S100 calcium binding protein A9 (calgranulin B)	MAPK	12.2	203535_AT
SCAMP1	secretory carrier membrane protein 1		6.2	206668_S_AT
SCAP2	src family associated phosphoprotein 2	Cell C.	6.2	216899_S_AT
SELL	selectin L (lymphocyte adhesion molecule 1)	MAPK	12.2	204563_AT
SEPT2	septin 2		6.2	200778_S_AT
SH3BP5	SH3-domain binding protein 5 (BTK-associated)	MAPK	3.2	201811_X_AT
SLA	Src-like-adaptor		13.2	203761_AT
SLC3A2	solute carrier family 3 (activators of dibasic and neutral amino acid transport), member 2		3.2	200924_S_AT
SNAP23	synaptosomal-associated protein, 23kDa		6.2	214544_S_AT
SOD2	superoxide dismutase 2, mitochondrial	MAPK	12.2	215223_S_AT
STK17A	serine/threonine kinase 17a (apoptosis-inducing)	MAPK	3.2	202693_S_AT
TALDO1	transaldolase 1		3.2	201463_S_AT
TLK1	tousled-like kinase 1		3.2	202606_S_AT
TLR4	toll-like receptor 4	Cell C.	6.2,13.2	221060_S_AT
TNFAIP3	tumor necrosis factor, alpha-induced protein 3		3.2, 12.2	202644_S_AT
TNFAIP6	tumor necrosis factor, alpha-induced protein 6		12.2	206026_S_AT
TPP1	tripeptidyl peptidase I		6.2	214196_S_AT
TSC22D1	TSC22 domain family, member 1		12.2	215111_S_AT
TXN	thioredoxin	MAPK	3.2	208864_S_AT
TXNDC	thioredoxin domain containing	Cell C.	6.2	208097_S_AT
TXNIP	thioredoxin interacting protein		6.2,13.2	201008_S_AT
TXNRD1	thioredoxin reductase 1		3.2	201266_AT
UCP2	uncoupling protein 2 (mitochondrial, proton carrier)		13.2	208997_S_AT
VAMP3	vesicle-associated membrane protein 3 (cellubrevin)		6.2	201337_S_AT
YWHAZ	tyrosine 3-monooxygenase/tryptophan 5-monooxygenase activation protein, zeta polypeptide		6.2	200641_S_AT
ZNFN1A1	zinc finger protein, subfamily 1A, 1 (Ikaros)		6.2	205039_S_AT

Table 4 - Regulatory sequences genes

SYMBOL	NAME	Pathway	CLU	Prob
ABCA1	ATP-binding cassette, sub-family A (ABC1), member 1		4.1	203505
ABCG1	ATP-binding cassette, sub-family G (WHITE), member 1		10.1	204567_S
ACAT2	acetyl-Coenzyme A acetyltransferase 2 (acetoacetyl Coenzyme A thiolase)		11.2	209608_S
ADM	adrenomedullin	TP53	14.1	202912
ALDH1A1	aldehyde dehydrogenase 1 family, member A1		4.1, 11.2	212224
ALDH2	aldehyde dehydrogenase 2 family (mitochondrial)		4.1	201425
ALOX5AP	arachidonate 5-lipoxygenase-activating protein		10.1	204174
ARTS-1	type 1 tumor necrosis factor receptor shedding aminopeptidase regulator		11.2	210385_S
C3AR1	complement component 3a receptor 1		4.1, 14.1	209906
CALR	calreticulin		10.1	214315_X
CCND2	cyclin D2	JUN/FOS, TP53	10.1, 14.1	200953_S
CDC42	cell division cycle 42 (GTP binding protein, 25kDa)	TP53	14.1	208727_S
CPM	carboxypeptidase M		11.2	206100
CREM	cAMP responsive element modulator	JUN/FOS	10.1	207630_S
CTSK	cathepsin K (pycnodysostosis)		4.1	202450_S
CTSL	cathepsin L		14.1	202087_S
CXCL1	chemokine (C-X-C motif) ligand 1 (melanoma growth stimulating activity, alpha)	JUN/FOS	10.1, 14.1	209774_X
CXCR4	chemokine (C-X-C motif) receptor 4		4.1	211919_S
CYP51A1	cytochrome P450, family 51, subfamily A, polypeptide 1		11.2	216607_S
EBP	emopamil binding protein (sterol isomerase)		11.2	202735
FBP1	fructose-1,6-bisphosphatase 1		10.1	209696
FDFT1	farnesyl-diphosphate farnesyltransferase 1		11.2	208647
FYB	FYN binding protein (FYB-120/130)		4.1	211795_S
G0S2	G0/G1switch 2		14.1	213524_S
G1P2	interferon, alpha-inducible protein (clone IFI-15K)		4.1	205483_S
GADD45A	growth arrest and DNA-damage-inducible, alpha	JUN/FOS, TP53	10.1, 14.1	203725
GCH1	GTP cyclohydrolase 1 (dopa-responsive dystonia)	TP53	14.1	204224_S
GGH	gamma-glutamyl hydrolase (conjugase, folylpolygammaglutamyl hydrolase)		11.2	203560
GM2A	GM2 ganglioside activator		4.1	212737
HLA-DMB	major histocompatibility complex, class II, DM beta		4.1	203932
HLA-DQA2	major histocompatibility complex, class II, DQ alpha 2		4.1, 14.1	212671_S
HLA-DQB1	major histocompatibility complex, class II, DQ beta 1		4.1	212998_X
HMGCR	3-hydroxy-3-methylglutaryl-Coenzyme A reductase		11.2	202540_S
HPSE	heparanase		14.1	219403_S
HSPA1B	heat shock 70kDa protein 1B		11.2	200800_S
IER3	immediate early response 3	TP53	14.1	201631_S
IL1RN	interleukin 1 receptor antagonist	JUN/FOS	10.1	212659_S
INSIG1	insulin induced gene 1		11.2	201625_S
JUN	v-jun sarcoma virus 17 oncogene homolog (avian)	JUN/FOS	4.1, 10.1	201466_S
LCP2	lymphocyte cytosolic protein 2 (SH2 domain containing leukocyte protein of 76kDa)		14.1	205269
LDLR	low density lipoprotein receptor (familial hypercholesterolemia)		11.2	202068_S
LOC440607	Fc-gamma receptor I B2		10.1	214511_X
LYZ	lysozyme (renal amyloidosis)		11.2, 14.1	213975_S
MAPK13	mitogen-activated protein kinase 13	JUN/FOS	10.1	210058
MARCKS	myristoylated alanine-rich protein kinase C substrate	JUN/FOS	4.1, 14.1	201670_S
MMP14	matrix metallopeptidase 14 (membrane-inserted)		10.1	160020
NISCH	nischarin		4.1	201591_S
NP	nucleoside phosphorylase		14.1	201695_S
PDGFC	platelet derived growth factor C		4.1	218718
PFKFB3	6-phosphofructo-2-kinase/fructose-2,6-biphosphatase 3		14.1	202464_S
PHLDA1	pleckstrin homology-like domain, family A, member 1		14.1	217996
PIM1	pim-1 oncogene	JUN/FOS, TP53	10.1, 14.1	209193
PLAU	plasminogen activator, urokinase	JUN/FOS	10.1	211668_S
PROCR	protein C receptor, endothelial (EPCR)		14.1	203650
PTGS1	prostaglandin-endoperoxide synthase 1 (prostaglandin G/H synthase and cyclooxygenase)		11.2	215813_S
RALA	v-ral simian leukemia viral oncogene homolog A (ras related)		10.1	214435_X
RDX	radixin		4.1	212397
RPS6KA4	ribosomal protein S6 kinase, 90kDa, polypeptide 4		11.2	204632
S100A12	S100 calcium binding protein A12 (calgranulin C)		14.1	205863

continued on next page

continued from previous page

SYMBOL	NAME	Pathway	CLU	ProbeSet
S100A8	S100 calcium binding protein A8 (calgranulin A)	JUN/FOS	10.1	202917_S_AT
SHMT2	serine hydroxymethyltransferase 2 (mitochondrial)		10.1	214437_S_AT
SLC11A1	solute carrier family 11 (proton-coupled divalent metal ion transporters), member 1		10.1	210423_S_AT
SOD2	superoxide dismutase 2, mitochondrial	JUN/FOS	4.1	215223_S_AT
SPINT2	serine peptidase inhibitor, Kunitz type, 2		14.1	210715_S_AT
SQLE	squalene epoxidase		11.2	209218_AT
TNFAIP6	tumor necrosis factor, alpha-induced protein 6		14.1	206026_S_AT
TRAPPC2	trafficking protein particle complex 2		4.1	209751_S_AT
TRIB3	tribbles homolog 3 (Drosophila)		14.1	218145_AT
UGCG	UDP-glucose ceramide glucosyltransferase		4.1,10.1	204881_S_AT

Table 5 - Differentiation and cell cycle genes

SYMBOL	NAME	Pathway	CLU	ProbeSet
AGA	aspartylglucosaminidase	TP53	11.1	204333_S_AT
ALCAM	activated leukocyte cell adhesion molecule		12.1	201951_AT
ALOX5	arachidonate 5-lipoxygenase		11.1	204446_S_AT
APP	amyloid beta (A4) precursor protein (peptidase nexin-II, Alzheimer disease)	TP53	5.2	214953_S_AT
ATP1B1	ATPase, Na+/K+ transporting, beta 1 polypeptide		12.1	201242_S_AT
CD44	CD44 antigen (homing function and Indian blood group system)	TP53	11.1	210916_S_AT
CDKN1A	cyclin-dependent kinase inhibitor 1A (p21, Cip1)	TP53	11.1	202284_S_AT
CSPG2	chondroitin sulfate proteoglycan 2 (versican)	TP53	5.2	221731_X_AT
CTNNB1	catenin (cadherin-associated protein), beta 1, 88kDa		11.1	201533_AT
CYP51A1	cytochrome P450, family 51, subfamily A, polypeptide 1	TP53	12.1	216607_S_AT
DUSP6	dual specificity phosphatase 6	TP53	5.2	208892_S_AT
DUT	dUTP pyrophosphatase		5.2,11.1	209932_S_AT
EIF2AK2	eukaryotic translation initiation factor 2-alpha kinase 2	TP53	5.2	204211_X_AT
EPRS	glutamyl-prolyl-tRNA synthetase		12.1	200842_S_AT
EREG	epiregulin		11.1	205767_AT
F8	coagulation factor VIII, procoagulant component (hemophilia A)		5.2	205756_S_AT
FCGR1A	Fc fragment of IgG, high affinity Ia, receptor (CD64)		5.2	216950_S_AT
FCGR3A	Fc fragment of IgG, low affinity IIIa, receptor (CD16a)		5.2	204007_AT
FYN	FYN oncogene related to SRC, FGR, YES		5.2	210105_S_AT
GCLC	glutamate-cysteine ligase, catalytic subunit		12.1	202923_S_AT
GGH	gamma-glutamyl hydrolase (conjugase, folylpolygammaglutamyl hydrolase)		5.2, 12.1	203560_AT
GSN	gelsolin (amyloidosis, Finnish type)	TP53	12.1	200696_S_AT
HMGB2	high-mobility group box 2	TP53	5.2	208808_S_AT
HMGB3	high-mobility group box 3		5.2, 11.1	203744_AT
HMGCR	3-hydroxy-3-methylglutaryl-Coenzyme A reductase	TP53	12.1	202540_S_AT
IL1RN	interleukin 1 receptor antagonist	TP53	12.1	212659_S_AT
ITGA4	integrin, alpha 4 (antigen CD49D, alpha 4 subunit of VLA-4 receptor)		5.2	205885_S_AT
LDLR	low density lipoprotein receptor (familial hypercholesterolemia)		5.2	202068_S_AT
LMNB1	lamin B1		5.2	203276_AT
LYZ	lysozyme (renal amyloidosis)		5.2	213975_S_AT
MCM5	MCM5 minichromosome maintenance deficient 5, cell division cycle 46 (S. cerevisiae)		5.2	216237_S_AT
NME1	non-metastatic cells 1, protein (NM23A) expressed in	TP53	12.1	201577_AT
PCNA	proliferating cell nuclear antigen	TP53	5.2,12.1	201202_AT
PDCD4	programmed cell death 4 (neoplastic transformation inhibitor)		5.2	212593_S_AT
PICALM	phosphatidylinositol binding clathrin assembly protein		11.1	212511_AT
PPP1R15A	protein phosphatase 1, regulatory (inhibitor) subunit 15A	TP53	11.1	37028_AT
PRKCA	protein kinase C, alpha	TP53	5.2	213093_AT
RRM1	ribonucleotide reductase M1 polypeptide		5.2	201477_S_AT
RUNX3	runt-related transcription factor 3	TP53	11.1	204198_S_AT
SELL	selectin L (lymphocyte adhesion molecule 1)		5.2	204563_AT
SLA	Src-like-adaptor		12.1	203761_AT
SLC7A1	solute carrier family 7 (cationic amino acid transporter, y+ system), member 1		12.1	212295_S_AT
SMARCA3	SWI/SNF related, matrix associated, actin dependent regulator of chromatin, subfamily a, member 3	TP53	5.2	202983_AT
SMC4L1	SMC4 structural maintenance of chromosomes 4-like 1 (yeast)		11.1	201664_AT
SOX4	SRY (sex determining region Y)-box 4		11.1	201417_AT
SPTBN1	spectrin, beta, non-erythrocytic 1		5.2	212071_S_AT
SRD5A1	steroid-5-alpha-reductase, alpha polypeptide 1 (3-oxo-5 alpha-steroid delta 4-dehydrogenase alpha 1)	TP53	11.1	204675_AT
TFDP1	transcription factor Dp-1		5.2	212330_AT

Table 6 - Survival/Apoptosis genes

SYMBOL	NAME	Pathway	CLU	ProbeSet
ADAM17	ADAM metallopeptidase domain 17 (tumor necrosis factor, alpha, converting enzyme)	TP53, BAX	13.1	205746_S_AT
ALOX5	arachidonate 5-lipoxygenase	BAX, FAS	3.1, 9.2	204446_S_AT
ALOX5AP	arachidonate 5-lipoxygenase-activating protein	BAX	3.1	204174_AT
ATF3	activating transcription factor 3	TP53, BAX	8.1	202672_S_AT
BAX	BCL2-associated X protein	TP53	3.1, 8.1,13.1	211833_S_AT
BCL2A1	BCL2-related protein A1		4.2	205681_AT
BTG1	B-cell translocation gene 1, anti-proliferative		13.1	200920_S_AT
C1QR1	complement component 1, q subcomponent, receptor 1	CALR	4.2	202878_S_AT
CACYBP	calcyclin binding protein		2.1	210691_S_AT
CALR	calreticulin		4.2	214315_X_AT
CCL3	chemokine (C-C motif) ligand 3	BAX	6.1 8.1	205114_S_AT
CCND2	cyclin D2		4.2	200953_S_AT
CD36	CD36 antigen (collagen type I receptor, thrombospondin receptor)	CALR	4.2	209555_S_AT
CD44	CD44 antigen (homing function and Indian blood group system)	TP53, FAS	9.2	204490_S_AT
CD83	CD83 antigen (activated B lymphocytes, immunoglobulin superfamily)	BAX	13.1	204440_AT
CHMP5	chromatin modifying protein 5		2.1	219356_AT
CSPG2	chondroitin sulfate proteoglycan 2 (versican)	TP53, FAS	9.2	221731_X_AT
CTSD	cathepsin D (lysosomal aspartyl peptidase)	TP53, EAX	3.1	200766_AT
CXCL1	chemokine (C-X-C motif) ligand 1 (melanoma growth stimulating activity, alpha)	BAX	6.1 8.1	204470_AT
CXCR4	chemokine (C-X-C motif) receptor 4	BAX	3.1,13.1	217028_AT
CYP51A1	cytochrome P450, family 51 subfamily A, polypeptide 1	TP53	2.1	202314_AT
DNM2	dynamin 2		4.2	202253_S_AT
DNTTIP2	deoxynucleotidyltransferase, terminal, interacting protein 2		6.1	202776_AT
DUSP1	dual specificity phosphatase 1	TP53, BAX	13.1	201041_S_AT
EGR2	early growth response 2 (Krox-20 homolog, Drosophila)	BAX	8.1,13.1	205249_AT
EIF5B	eukaryotic translation initiation factor 5B	TP53	8.1	201027_S_AT
ERCC1	excision repair cross-complementing rodent repair deficiency, complementation group 1		4.2	203719_AT
F8	coagulation factor VIII, procoagulant component (hemophilia A)		9.2	205756_S_AT
FAS	Fas (TNF receptor superfamily, member 6)	TP53	9.2	204780_S_AT
FCGR1A	Fc fragment of IgG, high affinity Ia, receptor (CD64)		3.1	216950_S_AT
FLJ22386	leucine zipper domain protein		13.1	218394_AT
FOXO1A	forkhead box O1A (rhabdomyosarcoma)	BAX	4.2, 6.1, 13.1	202724_S_AT
FYB	FYN binding protein (FYB-120/130)	BAX	8.1	211795_S_AT
FYN	FYN oncogene related to SRC, FGR, YES	BAX	13.1	210105_S_AT
GADD45A	growth arrest and DNA-damage-inducible, alpha	TP53, BAX	3.1, 6.1	203725_AT
GRB10	growth factor receptor-bound protein 10		9.2	209409_AT
HEBP2	heme binding protein 2	FAS	9.2	203430_AT
HLA-DQA1	major histocompatibility complex, class II, DQ alpha 1	TP53	2.1, 3.1, 13.1	213831_AT
IER3	immediate early response 3	P53	6.1	201631_S_AT
IGFBP7	insulin-like growth factor binding protein 7		2.1	201163_S_AT
IL1RN	interleukin 1 receptor antagonist	TP53, BAX	8.1,13.1	212659_S_AT
ING1	inhibitor of growth family, member 1	TP53, BAX	13.1	208415_X_AT
IRS2	insulin receptor substrate 2	BAX	13.1	209185_S_AT
ITGAL	integrin, alpha L (antigen CD11A (p180), lymphocyte function-associated antigen 1; alpha polypeptide)	BAX	13.1	213475_S_AT
JAG1	jagged 1 (Alagille syndrome)	P53	6.1	209099_X_AT
LAMP1	lysosomal-associated membrane protein 1	BAX	3.1	201551_S_AT
LNK	lymphocyte adaptor protein		4.2	203320_AT
LRMP	lymphoid-restricted membrane protein		3.1	35974_AT
LY75	lymphocyte antigen 75	FAS	9.2	205668_AT
MAP2K3	mitogen-activated protein kinase kinase 3		4.2	215498_S_AT
MAP3K5	mitogen-activated protein kinase kinase 5		6.1	203836_S_AT
MCL1	myeloid cell leukemia sequence 1 (BCL2-related)		4.2	200798_X_AT
NDUFA5	NADH dehydrogenase (ubiquinone) 1 alpha subcomplex, 5, 13kDa	TP53	2.1	201304_AT
NEDD8	neural precursor cell expressed, developmentally down-regulated 8		2.1	201840_AT
NFKB2	nuclear factor of kappa light polypeptide gene enhancer in B-cells 2 (p49/p100)	TP53, BAX	13.1	207535_S_AT
NME1	non-metastatic cells 1, protein (NM23A) expressed in	CALR	4.2	201577_AT
OLR1	oxidised low density lipoprotein (lectin-like) receptor 1		13.1	210004_AT
PCBP2	poly(rC) binding protein 2		13.1	213263_S_AT

continued on next page

continued from previous page

SYMBOL	NAME	Pathway	CLU	ProbeSet
PCNA	proliferating cell nuclear antigen	TP53	2.1	201202_AT
PDE4B	phosphodiesterase 4B, cAMP-specific (phosphodiesterase E4 dunce homolog, Drosophila)		13.1	203708_AT
PER2	period homolog 2 (Drosophila)		6.1	205251_AT
PLEK	pleckstrin		13.1	203470_S_AT
PPP1R15A	protein phosphatase 1, regulatory (inhibitor) subunit 15A	TP53, BAX	13.1	37028_AT
PRKACB	protein kinase, cAMP-dependent, catalytic, beta		3.1	202741_AT
PRKCB1	protein kinase C, beta 1		4.2	209685_S_AT
PROCR	protein C receptor, endothelial (EPCR)	CALR	4.2	203650_AT
PSMB8	proteasome (prosome, macropain) subunit, beta type, 8 (large multifunctional peptidase 7)		9.2	209040_S_AT
RAD23B	RAD23 homolog B (S. cerevisiae)		2.1	201222_S_AT
RALGDS	ral guanine nucleotide dissociation stimulator	FAS	9.2	209050_S_AT
REL	v-rel reticuloendotheliosis viral oncogene homolog (avian)	BAX	13.1	206036_S_AT
RNASE2	ribonuclease, RNase A family, 2 (liver, eosinophil-derived neurotoxin)		9.2	206111_AT
RSL1D1	ribosomal L1 domain containing 1		2.1, 4.2	212018_S_AT
S100A8	S100 calcium binding protein A8 (calgranulin A)	TP53, FAS	9.2	202917_S_AT
S100A9	S100 calcium binding protein A9 (calgranulin B)	TP53, FAS	9.2	203535_AT
SERBP1	SERPINE1 mRNA binding protein 1		9.2	210466_S_AT
SFRS5	splicing factor, arginine/serine-rich 5		13.1	212266_S_AT
SLC11A1	solute carrier family 11 (proton-coupled divalent metal ion transporters), member 1	CALR	4.2,9.2	210423_S_AT
SOD2	superoxide dismutase 2, mitochondrial	TP53	2.1	216841_S_AT
SOX4	SRY (sex determining region Y)-box 4		13.1	201417_AT
SPTBN1	spectrin, beta, non-erythrocytic 1		13.1	212071_S_AT
STEAP3	STEAP family member 3		9.2	218424_S_AT
SUB1	SUB1 homolog (S. cerevisiae)		2.1	214512_S_AT
TANK	TRAF family member-associated NFKB activator		2.1	209451_AT
TGFB1	transforming growth factor, beta 1 (Camurati-Engelmann disease)	TP53, BAX	13.1	203085_S_AT
TNFSF10	tumor necrosis factor (ligand) superfamily, member 10	TP53, BAX	2.1, 3.1, 6.1	202688_AT
TNFSF13	tumor necrosis factor (ligand) superfamily, member 13	BAX	3.1	210314_X_AT
TRIB3	tribbles homolog 3 (Drosophila)		4.2	218145_AT
UBE1C	ubiquitin-activating enzyme E1C (UBA3 homolog, yeast)		2.1	209115_AT
VIL2	villin 2 (ezrin)	TP53, FAS	9.2	208623_S_AT
WARS	tryptophanyl-tRNA synthetase		6.1	200629_AT

B. Intronic miRNAs

B.1. Intronic miRNAs and host genes

Host Gene	miRNA	locus
1110001A07Rik	mmu-miR-301	intron
2010111I01Rik	mmu-miR-24-1	intron
2010209O12Rik	mmu-miR-671	exon
2610203C20Rik	mmu-miR-125b-1	intron
6230410P16Rik	mmu-miR-135a-1	antisense
Aatk	mmu-miR-338	intron
Acadvl	mmu-miR-324	antisense
Akt1s1	mmu-miR-707	intron
Ank1	mmu-miR-486	intron
Arpp21	mmu-miR-128b	intron
Arrb1	mmu-miR-326	intron
Astn1	mmu-miR-488	intron
Atp5b	mmu-miR-677	intron
Bcl7c	mmu-miR-762	antisense
Calcr	mmu-miR-489	intron
Cdkl1	mmu-miR-681	intron
Chm	mmu-miR-361	intron
Col27a1	mmu-miR-455	intron
Col7a1	mmu-miR-711	intron
Copz1	mmu-miR-148b	intron
Copz2	mmu-miR-152	intron
Ctdsp1	mmu-miR-26b	intron
Ctdspl	mmu-miR-26a-1	intron
Cutl1	mmu-miR-721	intron
D16H22S680E	mmu-miR-185	intron
Dnaja1	mmu-miR-207	intron
Dnm1	mmu-miR-199b	antisense
Dnm2	mmu-miR-199a-1	antisense
Dnm3	mmu-miR-199a-2	antisense

Dnm3os	mmu-miR-199a-2	intron
Dusp19	mmu-miR-684-1	antisense
Dvl2	mmu-miR-324	intron
Eda	mmu-miR-676	intron
Egfl7	mmu-miR-126	intron
Elmo3	mmu-miR-328	antisense
Evl	mmu-miR-342	intron
Ftl1	mmu-miR-692-2	exon
Gabre	mmu-miR-452	intron
Gpc1	mmu-miR-149	intron
Gpc3	mmu-miR-717	intron
Grid1	mmu-miR-346	intron
Grik3	mmu-miR-692-2	intron
Grm8	mmu-miR-592	intron
H19	mmu-miR-675	intron/exon
Hnrpk	mmu-miR-7-1	intron
Hoxc5	mmu-miR-615	intron
Htr2c	mmu-miR-764	intron
Huwe1	mmu-miR-98	intron
Iars2	mmu-miR-215	antisense
Igf2	mmu-miR-483	intron
Inpp5b	mmu-miR-698	exon
Irak1	mmu-miR-718	exon
Lpp	mmu-miR-28	intron
Map2k4	mmu-miR-744	intron
Mcm7	mmu-miR-25	intron
Mest	mmu-miR-335	intron
Mib1	mmu-miR-1-2	antisense
Myh6	mmu-miR-208	intron
Nfyc	mmu-miR-30c-1	intron
Nr6a1	mmu-miR-181b-2	antisense
Nrd1	mmu-miR-761	intron
Nupl1	mmu-miR-719	exon
Pank1	mmu-miR-107	intron
Pank3	mmu-miR-103-1	intron
Pde2a	mmu-miR-139	intron
Pdia4	mmu-miR-704	exon
Plod3	mmu-miR-702	exon
Ppargc1b	mmu-miR-378	intron
Prmt2	mmu-miR-678	exon
Psmb5	mmu-miR-686	exon

Ptk2	mmu-miR-151	intron
Ptprn2	mmu-miR-153	intron
R3hdm1	mmu-miR-128a	intron
Rab11fip5	mmu-miR-705	exon
Rb1	mmu-miR-687	intron
Rcl1	mmu-miR-101b	intron
Rfx1	mmu-miR-709	intron
Rnf130	mmu-miR-340	intron
Robo2	mmu-miR-691	intron
Rtl1	mmu-miR-434	antisense
Sf3a3	mmu-miR-697	intron
Sfmbt2	mmu-miR-297b	intron
Slit2	mmu-miR-218-1	intron
Slit3	mmu-miR-218-2	intron
Smc4	mmu-miR-16-2	intron
Srebf2	mmu-miR-33	intron
Tln2	mmu-miR-190	intron
Tmem49	mmu-miR-21	3' UTR/ exon
Trpm1	mmu-miR-211	intron
Trpm3	mmu-miR-204	intron
Ttc28	mmu-miR-701	intron
Ttll10	mmu-miR-429	intron
Wdr82	mmu-let-7g	intron
Wnk1	mmu-miR-706	intron
Wwp2	mmu-miR-140	intron
Xpo5	mmu-miR-693	intron
Zc3h7a	mmu-miR-689-2	antisense
Zranb2	mmu-miR-186	intron

B.2. MicroRNA host gene cluster

Each row defines one cluster.

Somitogenesis dataset:
1110001a07rik, Chm, Copz1, Dnm1, Gpc1, Iars2, Mcm7, Nupl1, Pank1, Prmt2, Sf3a3, Xpo5, Zranb2

2010111i01rik, Calcr, Chm, Ctdsp1, Cutl1, Dnm1, Dnm3, Dnm3os, Dusp19, Dvl2, Evl, Gabre, Igf2, Mest, Rab11fip5, Rb1, Slit2, Smc4, Srebf2, Ttc28, Wwp2

6230410p16rik, Arrb1, Chm, Elmo3, Htr2c, Irak1, Pde2a

Chm, Psmb5

Ctdspl, Igf2

Dnaja1, Rcl1, Tmem49

Neurite outgrowth dataset:
1110001a07rik, Pank1, Slit2, Ttc28,

Aatk, Bcl7c, Copz2, Ctdsp1, H19, Igf2, Irak1, Lpp, Plod3, Rnf130

Acadvl, Igf2

Astn1, Atp5b, Chm, Copz1, Dnaja1, Dnm1, Evl, Igf2, Map2k4, Nrd1, Nupl1, Psmb5, Rb1, Rcl1, Sf3a3, Tmem49, Zc3h7a

Stem cell development dataset:
Acadvl, Elmo3, Gpc3, H19, Igf2, Lpp, Plod3, Rab11fip5, Rnf130

Dnaja1, Hnrpk, Igf2, Nfyc, Psmb5, Rcl1, Sf3a3, Xpo5

Igf2, Wnk1

B.3. Functional similarity

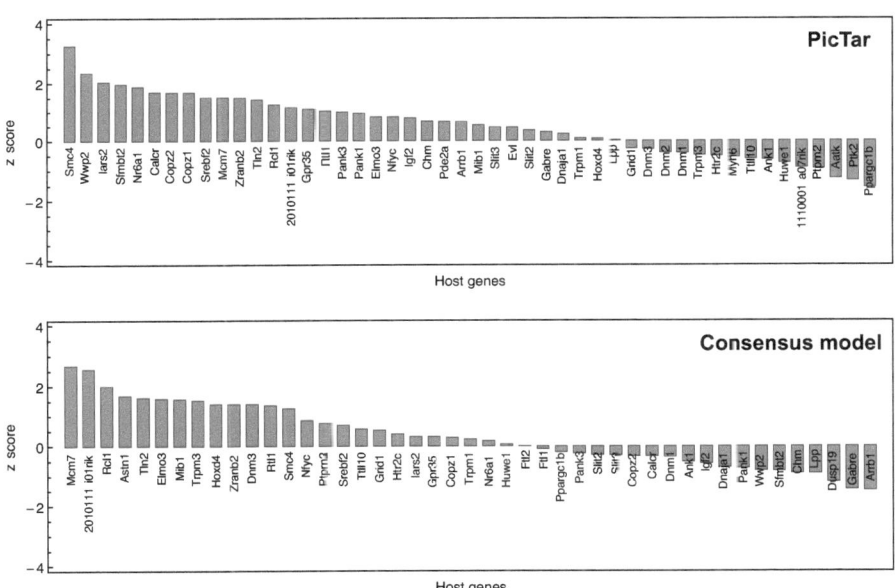

Figure B.1.: Functional similarity of host and target gene sets as predicted by Pictar and the consensus model. Z scores for all annotated host genes.

Acknowledgements

So now at the end of this work I'd like to express my greatest gratitude to all the people helped and accompanied me during the last years, months and weeks up to today. I could never have done this work without all the help, support, guidance and efforts of the following people:

- Prof. Dr. Elmar W. Lang. He was the first supervisor of my work and had also accompanied me during my diploma thesis and parts of my undergraduate studies. He strongly supported me in all my interests not only in the field of information theoretical based methods for microarray data analysis but also in other fields including machine learning and computational intelligence. Furthermore he gave me the opportunity to participate at an international conference and several visits to the work group of Pedro Gomez Vilda in Madrid. Moreover, he always willingly took the time to listen to all my questions, doubts and concerns.

- Prof. Dr. Dr. Fabian Theis. Halfway through my work he gave me the opportunity to slightly modify the focus of my work towards a more systems biology oriented focus and still continue with my already commenced projects. He also supported and still supports me in my ideas. He further made it possible to me to visit several interesting conferences and workshops. At any time, he readily listened and helped me to overcome all the problems I addressed to him. Finally he gives me the chance to carry on with my exciting running projects.

- Prof. Dr. Schmitz. I started my work under his guidance. During my time at the 'Klinikum Regensburg', he made several datasets available and guided my analyses at his institute. Furthermore, he taught me to self-confidently continue and defend my work.

- Prof. Dr. Pedro Gomez Vilda. During my visits to Madrid he was a pleasant host.

We had several interesting discussions and he not only supplied intersting ideas but also various impressive informations about Spain, its culture and history.

- Dr. PD Thomas Langmann and Prof. Dr Charalampos Aslanidis. They supported me in all my biological questions and guided my work during my studies at the 'Klinikum Regensburg'.

- Dr. Peter Ugocsai. He was not only my office colleague in Regensburg, but also a readily helpful and very friendly fellow in many projects. I also extremely enjoyed our football plays.

- My further co-operation partners. Without them it would not have been possible to get any of my work published. In particular there were Margot Grandl, Peter Gruber, Daniela Herold, Jan Krumsiek, Carsten Marr, Christoph Möhle, Reinhard Schachtner, Erik Seibold and Wolf Splettstößer.

- Lena, Dominik W., Daniel and Carsten for proofreading my manuscript and many useful comments.

- My fellows here at the CMB and IBIS: Andreas, Elisabeth, Florian Bl., Florian Bü., Gabi, Giovani, Gitti, Konrad and Mara. Thanks for interesting projects, discussions, suggestions and table football matches.

- My old fellows in Regensburg at the CIML: Dominik S., Hans, Harold, Hermann, Ingo, Kurt and Mathias. Thanks for discussions, suggestions, tips on linux and blithely barbecues.

- My old fellows at the Institute for Klin. Chem. in Regensburg: Gerhard, Katharina, Kerstin, Markus, Max, Richard, Susanne and Wolfgang. I had a diversified time there but you made the hard ones livable.

- A million thanks to all my friends. There are so many of you, it is impossible to me to mention you all. Thank you for accompany me through all ups and downs, boredom, parties, trips, sports and of course music.

- My family, Mom, Dad, Sebastian, Tobias, Franziska, Julian and Samuel. Thank you for keep me always grounded and supporting and accompany me in all my decisions and plans I had (and still have), the wise and erroneous.

Bibliography

I. Affymetrix. GeneChip® Expression Analysis Technical Manual. *Santa Clara (CA): Affymetrix*, 2001.

B. Alberts, D. Bray, J. Lewis, M. Raff, K. Roberts, and J. Watson. *Molecular Biology of the Cell*. 2002.

T. D. Allen, J. M. Cronshaw, S. Bagley, E. Kiseleva, and M. W. Goldberg. The nuclear pore complex: mediator of translocation between nucleus and cytoplasm. *J Cell Sci*, 113 (Pt 10): 1651–1659, May 2000.

D. B. Allison, X. Cui, G. P. Page, and M. Sabripour. Microarray data analysis: from disarray to consolidation and consensus. *Nat Rev Genet*, 7(1):55–65, Jan 2006.

D. Allocco, I. Kohane, and A. Butte. Quantifying the relationship between co-expression, co-regulation and gene function. *BMC Bioinformatics*, 5:18, Feb 2004.

V. Ambros. The evolution of our thinking about micrornas. *Nat Med*, 14(10):1036–1040, Oct 2008.

H. Andersson, B. Hartmanová, P. Rydén, L. Noppa, L. Näslund and A. Sjöstedt. A microarray analysis of the murine macrophage response to infection with Francisella tularensis LVS. *J Med Microbiol*, 55(Pt 8):1023–1033, Aug 2006.

S. V. Anisimov. Serial Analysis of Gene Expression (SAGE): 13 years of application in research. *Curr Pharm Biotechnol*, 9(5):333–350, Oct 2008.

S. Arnaudeau, M. Frieden, K. Nakamura, C. Castelbou, M. Michalak, and N. Demaurex. Calreticulin differentially modulates calcium uptake and release in the endoplasmic reticulum and mitochondria. *Journal of Biological Chemistry*, 277:46696–46705, 2002.

M. Ashburner, C. Ball, J. Blake, D. Botstein, H. Butler, J. Cherry, A. Davis, K. Dolinski, S. Dwight, J. Eppig, et al. Gene Ontology: tool for the unification of biology. *Nature Genetics*, 25:25–29, 2000.

R. Autio, S. Kilpinen, M. Saarela, O. Kallioniemi, S. Hautaniemi, and J. Astola. Comparison of Affymetrix data normalization methods using 6,926 experiments across five array generations. *BMC Bioinformatics*, 10 Suppl 1:S24, 2009.

D. Baek, J. Villén, C. Shin, F. D. Camargo, S. P. Gygi, and D. P. Bartel. The impact of microRNAs on protein output. *Nature*, 455(7209):64–71, Sep 2008.

S. Barik. An intronic microRNA silences genes that are functionally antagonistic to its host gene. *Nucleic Acids Res*, 36(16):5232–5241, Sep 2008.

C. Barrandon, B. Spiluttini, and O. Bensaude. Non-coding RNAs regulating the transcriptional machinery. *Biol Cell*, 100(2):83–95, Feb 2008.

T. Barrett and R. Edgar. Gene Expression Omnibus (GEO): Microarray data storage, submission, retrieval, and analysis. *Methods in enzymology*, 411:352, 2006.

T. Barrett, D. B. Troup, S. E. Wilhite, P. Ledoux, D. Rudnev, C. Evangelista, I. F. Kim, A. Soboleva, M. Tomashevsky, and R. Edgar. NCBI GEO: mining tens of millions of expression profiles–database and tools update. *Nucleic Acids Res*, 35(Database issue):D760–D765, Jan 2007.

D. P. Bartel. MicroRNAs Genomics, Biogenesis, Mechanism, and Function. *Cell*, 116(2):281–297, 2004.

D. P. Bartel. MicroRNAs: target recognition and regulatory functions. *Cell*, 136(2):215–233, Jan 2009.

D. P. Bartel and C. Chen. Micromanagers of gene expression: the potentially widespread influence of metazoan microRNAs. *Nat Rev Genet*, 5(5):396–400, May 2004. doi: 10.1038/nrg1328. URL http://dx.doi.org/10.1038/nrg1328.

S. Baskerville and D. P. Bartel. Microarray profiling of microRNAs reveals frequent coexpression with neighboring miRNAs and host genes, 2005.

S. Becker, M. Warren, and S. Haskill. Colony-stimulating factor-induced monocyte survival and differentiation into macrophages in serum-free cultures. *The Journal of Immunology*, 139: 3703–3709, 1987.

G. BehreDagger, A. J. Whitmarsh, P. Matthew, Coghlanparallel, T. Hoang, L. C. Carpenterparallel, D.-E. ZhangDagger, R. J. Davis, and D. G. TenenDagger. c-Jun Is a JNK-independent Coactivator of the PU.1 Transcription Factor. *J Biol Chem*, 274:4939–4946, 1999.

E. Berezikov, W. Chung, J. Willis, E. Cuppen, and E. Lai. Mammalian Mirtron Genes. *Molecular Cell*, 28(2):328–336, 2007.

D. P. Berrar, W. Dubitzky, and M. Granzow. *A practical approach to microarray data analysis*. Kluwer academic publishers, 2003.

F. Besse and A. Ephrussi. Translational control of localized mRNAs: restricting protein synthesis in space and time. *Nat Rev Mol Cell Biol*, 9(12):971–980, Dec 2008.

D. Betel, M. Wilson, A. Gabow, D. S. Marks, and C. Sander. The microRNA.org resource: targets and expression. *Nucleic Acids Res*, 36(Database issue):D149–D153, Jan 2008.

B. Beutler. Inferences, questions and possibilities in Toll-like receptor signalling. *Nature*, 430: 257–263, 2004.

E. Birney, J. Stamatoyannopoulos, A. Dutta, R. Guigó, T. Gingeras, E. Margulies, Z. Weng, M. Snyder, E. Dermitzakis, J. Stamatoyannopoulos, et al. Identification and analysis of functional elements in 1% of the human genome by the ENCODE pilot project. *Nature*, 447(7146): 799–816, 2007.

N. H. Bishopric and K. A. Webster. Preventing Apoptosis With Thioredoxin. *Circ Res.*, 90: 1237–1239, 2002.

M. Bogoyevitch, I. Boehm, A. Oakley, A. Ketterman, and R. Barr. Targeting the JNK MAPK cascade for inhibition: basic science and therapeutic potential. *Biochim Biophys Acta.*, 1697: 89–101, 2004.

M. Böhm, K. Stadlthanner, P. Gruber, F. J. Theis, E. W. Lang, A. M. Tome, A. R. Teixeira, W. Gronwald, and H. R. Kalbitzer. On the use of simulated annealing to automatically assign decorrelated components in second-order blind source separation. *IEEE Trans Biomed Eng*, 53(5):810–820, May 2006.

J. Booth, W. Trimble, and S. Grinstein. Membrane dynamics in phagocytosis. *Semin Immunol.*, 13:357–64, 2001.

G. Borchert, W. Lanier, and B. Davidson. RNA polymerase III transcribes human microRNAs. *Nature Structural & Molecular Biology*, 13:1097–1101, 2006.

H. Branton and J. Kopp. TGF-beta and fibrosis. *Microbes Infect.*, 1:1349–65, 1999.

J. Brennecke, A. Stark, R. B. Russell, and S. M. Cohen. Principles of microRNA-target recognition. *PLoS Biol*, 3(3):e85, Mar 2005.

J.-P. Brunet, P. Tamayo, T. R. Golub, and J. P. Mesirov. Metagenes and molecular pattern discovery using matrix factorization. *Proc Natl Acad Sci U S A*, 101(12):4164–4169, Mar 2004.

N. Bushati and S. M. Cohen. MicroRNA functions. *Annu Rev Cell Dev Biol*, 23:175–205, 2007.

P. Callinan and A. Feinberg. The emerging science of epigenomics. *Human Molecular Genetics*, 15:R95, 2006.

I. G. Cannell, Y. W. Kong, and M. Bushell. How do microRNAs regulate gene expression? *Biochem Soc Trans*, 36(Pt 6):1224–1231, Dec 2008.

J. Cardoso and A. Souloumiac. Jacobi angles for simultaneous diagonalization. *SIAM Journal on Matrix Analysis and Applications*, 17(1):161–164, 1996.

J. Cardoso, A. Souloumiac, and T. Paris. Blind beamforming for non-Gaussian signals. In *IEE Proceedings F Radar and Signal Processing*, volume 140, pages 362–370, 1993.

P. Carninci. Non-coding RNA transcription: turning on neighbours. *Nat Cell Biol*, 10(9):1023–1024, Sep 2008.

P. Carninci and Y. Hayashizaki. Noncoding RNA transcription beyond annotated genes. *Curr Opin Genet Dev*, 17(2):139–144, Apr 2007.

P. Carninci, J. Yasuda, and Y. Hayashizaki. Multifaceted mammalian transcriptome. *Curr Opin Cell Biol*, 20(3):274–280, Jun 2008.

J. Carrington and V. Ambros. Role of MicroRNAs in Plant and Animal Development. *Science*, 301(5631):336–338, 2003.

M. K. Cathcart. Regulation of Superoxide Anion Production by NADPH Oxidase in Monocytes/Macrophages. *Arteriosclerosis, Thrombosis, and Vascular Biology*, 24:23–28, 2004.

S. Chan and F. Slack. And Now Introducing Mammalian Mirtrons. *Developmental Cell*, 13(5): 605–607, 2007.

P. Chiappetta, M. Roubaud, and B. Torresani. Blind source separation and the analysis of microarray data. *J Comput Biol.*, 11:1090–109, 2004.

A. Cichocki and S.-I. Amari. *Adaptive Blind Signal and Image Processing*. Wiley, 2002.

A. Cichocki, R. Zdunek, S. Choi, R. Plemmons, and S. Amari. Novel multi-layer non-negative tensor factorization with sparsity constraints. *Lecture Notes in Computer Science*, 4432:271, 2007.

P. Comon et al. Independent component analysis, a new concept. *Signal Processing*, 36(3): 287–314, 1994.

X. Cui and G. A. Churchill. Statistical tests for differential expression in cDNA microarray experiments. *Genome Biol*, 4(4):210, 2003.

B. R. Cullen. Nuclear RNA export. *J Cell Sci*, 116(Pt 4):587–597, Feb 2003.

B. R. Cullen. Transcription and processing of human microRNA precursors. *Mol Cell*, 16(6): 861–865, Dec 2004.

M. Desjardins. ER-mediated phagocytosis: a new membrane for new functions. *Nat Rev Immunol.*, 3:280–291, 2003.

J. G. Doench and P. A. Sharp. Specificity of microRNA target selection in translational repression. *Genes Dev*, 18(5):504–511, Mar 2004.

E. Dougherty, A. Datta, and C. Sima. Research issues in genomic signal processing. *IEEE Signal Processing Magazine*, 22(6):46–68, 2005.

S. Draghici, P. Khatri, A. L. Tarca, K. Amin, A. Done, C. Voichita, C. Georgescu, and R. Romero. A systems biology approach for pathway level analysis. *Genome Res*, 17(10):1537–1545, Oct 2007.

J. Dumic, S. Dabelic, and M. Flogel. Galectin-3: an open-ended story. *Biochim Biophys Acta.*, 1760:616–35, 2006.

B. Fabriek, C. Dijkstra, and T. van den Berg. The macrophage scavenger receptor CD163. *Immunobiology*, 210:153–60, 2005.

K. Farh, A. Grimson, C. Jan, B. Lewis, W. Johnston, L. Lim, C. Burge, and D. Bartel. The widespread impact of mammalian MicroRNAs on mRNA repression and evolution. *Science*, 310(5755):1817–1821, Dec 2005.

A. S. Flynt and E. C. Lai. Biological principles of microRNA-mediated regulation: shared themes amid diversity. *Nat Rev Genet*, 9(11):831–842, Nov 2008.

H. J. Forman and M. Torres. Reactive Oxygen Species and Cell Signaling; Respiratory Burst in Macrophage Signaling. *American Journal of Respiratory and Critical Care Medicine*, 166: S4–S8, 2002.

S. Fox, S. Haque, A. Lovibond, and T. Chambers. The possible role of TGF-beta-induced suppressors of cytokine signaling expression in osteoclast/macrophage lineage commitment in vitro. *Journal of Immunology*, 170:3679–87, 2003.

N. Friedman. Inferring cellular networks using probabilistic graphical models. *Science*, 303 (5659):799–805, Feb 2004.

V. Gangaraju and H. Lin. MicroRNAs: key regulators of stem cells. *Nat Rev Mol Cell Biol*, 10 (2):116–125, Feb 2009.

Genomatix. GenomatixSuite, München, Bayern/Deutschland: Genomatix Software GmbH, 2009. URL http://www.genomatix.de/.

M. B. Gerstein, C. Bruce, J. S. Rozowsky, D. Zheng, J. Du, J. C. Korbel, O. Emanuelsson, Z. D. Zhang, S. Weissman, and M. Snyder. What is a gene, post-ENCODE? History and updated definition. *Genome Res*, 17(6):669–681, Jun 2007.

G. Gill. Regulation of the initiation of eukaryotic transcription. *Essays Biochem*, 37:33–43, 2001.

S. Griffiths-Jones, R. Grocock, S. van Dongen, A. Bateman, and A. Enright. miRBase: microRNA sequences, targets and gene nomenclature. *Nucleic Acids Research*, 34:D140, 2006.

H. Grosshans and W. Filipowicz. Molecular biology: the expanding world of small RNAs. *Nature*, 451(7177):414–416, Jan 2008.

P. Gruber, K. Stadlthanner, M. Böhm, F. Theis, E. Lang, A. Tomé, A. Teixeira, C. Puntonet, and J. Górriz Saéz. Denoising using local projective subspace methods. *Neurocomputing*, 69 (13-15):1485–1501, 2006.

P. Gruber, H. Gutch, and F. Theis. Hierarchical extraction of independent subspaces of unknown dimensions. In *Independent Component Analysis and Signal Separation*, pages 259–266. Springer, 2009.

M. Habl, C. Bauer, C. Ziegaus, and E. W. Lang. *Perspectives in Neuroscience: Artificial Neural Networks in Medicine and Biology*, chapter Analyzing Brain Tumor Related EEG Signals With ICA Algorithms, pages 131 – 135. Springer Publishers Berlin, 2000.

A. Hackstadt and A. Hess. Filtering for Increased Power for Microarray Data Analysis. *BMC Bioinformatics*, 10(1):11, Jan 2009.

T. Henry and D. M. Monack. Activation of the inflammasome upon Francisella tularensis infection: interplay of innate immune pathways and virulence factors. *Cell Microbiol*, 9(11): 2543–2551, Nov 2007.

D. Herold, D. Lutter, R. Schachtner, A. M. Tomé, G. Schmitz, and E. W. Lang. Comparison of unsupervised and supervised gene selection methods. *Conf Proc IEEE Eng Med Biol Soc*, 2008:5212–5215, 2008.

S. Himes, D. Sester, T. Ravasi, S. Cronau, T. Sasmono, and D. Hume. The jnk are important for development and survival of macrophages. *Journal of Immunology*, 176:2219–2228, 2006.

J. Holmberg, D. L. Clarke, and J. Frisén. Regulation of repulsion versus adhesion by different splice forms of an Eph receptor. *Nature*, 408(6809):203–206, Nov 2000.

M. Houde, S. Bertholet, E. Gagnon, S. Brunet, G. Goyette, A. Laplante, M. F. Princiotta, P. Thibault, D. Sacks, and M. Desjardins. Phagosomes are competent organelles for antigen cross-presentation. *Nature*, 425:402–406, 2003.

G. Hutvágner, J. McLachlan, A. E. Pasquinelli, E. Bálint, T. Tuschl, and P. D. Zamore. A cellular function for the RNA-interference enzyme Dicer in the maturation of the let-7 small temporal RNA. *Science*, 293(5531):834–838, Aug 2001.

H. Hwang and J. Mendell. MicroRNAs in cell proliferation, cell death, and tumorigenesis. *British Journal of Cancer*, 94:776–780, 2006.

A. Hyvärinen. Fast and robust fixed-point algorithms for independent component analysis. *IEEE Trans Neural Netw*, 10(3):626–634, 1999.

A. Hyvärinen, J. Karhunen, and E. Oja. Independent Component Analysis. *John Wiley & Sons*, 2001.

A. Hyvärinen, J. Karhunen, and E. Oja. *Independent component analysis*. 2001.

N. Iglesias and F. Stutz. Regulation of mRNP dynamics along the export pathway. *FEBS Lett*, 582(14):1987–1996, Jun 2008.

R. A. Irizarry, B. M. Bolstad, F. Collin, L. M. Cope, B. Hobbs, and T. P. Speed. Summaries of Affymetrix GeneChip probe level data. *Nucleic Acids Res*, 31(4):e15, Feb 2003.

M. Kanehisa, M. Araki, S. Goto, M. Hattori, M. Hirakawa, M. Itoh, T. Katayama, S. Kawashima, S. Okuda, T. Tokimatsu, and Y. Yamanishi. KEGG for linking genomes to life and the environment. *Nucleic Acids Res*, 36(Database issue):D480–D484, Jan 2008.

E. S. Kawasaki. The end of the microarray Tower of Babel: will universal standards lead the way? *J Biomol Tech*, 17(3):200–206, Jul 2006.

I. R. Keck, F. J. Theis, P. Gruber, E. W. Lang, K. Specht, and C. G. Puntonet. 3D spatial analysis of fMRI data on a word perception task. In C. G. Puntonet and A. Prieto, editors, *Lecture Notes in Computer Science, LNCS 3195*, pages 977–984, Berlin, 2004. Springer Verlag.

M. Kertesz, N. Iovino, U. Unnerstall, U. Gaul, and E. Segal. The role of site accessibility in microRNA target recognition. *Nature Genetics*, 39(10):1278, 2007.

R. F. Ketting, S. E. Fischer, E. Bernstein, T. Sijen, G. J. Hannon, and R. H. Plasterk. Dicer functions in RNA interference and in synthesis of small RNA involved in developmental timing in C. elegans. *Genes Dev*, 15(20):2654–2659, Oct 2001.

Y. Kim and V. Kim. Processing of intronic microRNAs. *The EMBO Journal*, 26:775–783, 2007.

T. Kouzarides. Chromatin Modifications and Their Function. *Cell*, 128(4):693–705, 2007.

A. Krek, D. Grün, M. Poy, R. Wolf, L. Rosenberg, E. Epstein, P. MacMenamin, I. da Piedade, K. Gunsalus, M. Stoffel, et al. Combinatorial microRNA target predictions. *Nature Genetics*, 37:495–500, 2005.

J. Krumsiek, C. Friedel, and R. Zimmer. Procope–protein complex prediction and evaluation. *Bioinformatics*, 24(18):2115–2116, Sep 2008.

G. Kustermans, J. E. Benna, J. Piette, and S. Legrand-Poels. Perturbation of actin dynamics induces nf-κb activation in myelomonocytic cells through an nadph oxidase-dependent pathway. *Biochem J.*, 387:531–540, 2005.

M. Lagos-Quintana, R. Rauhut, W. Lendeckel, and T. Tuschl. Identification of Novel Genes Coding for Small Expressed RNAs, 2001.

M. Lagos-Quintana, R. Rauhut, J Meyer, A. Borkhardt, and T. Tuschl. New microRNAs from mouse and human. *RNA*, 9(2):175–179, Feb 2003.

T. Langmann, C. S. S. G. Morham, C. Honer, S. Heimerl, C. Moehle, and G. Schmitz. ZNF202 is inversely regulated with its target genes ABCA1 and apoE during macrophage differentiation and foam cell formation. *Lipid Res*, 44:968–977, 2003.

D. Lau, H. Mollnau, J. P. Eiserich, B. A. Freeman, A. Daiber, U. M. Gehling, J. Brümmer, V. Rudolph, T. Münzel, T. Heitzer, T. Meinertz, and S. Baldus. Myeloperoxidase mediates neutrophil activation by association with CD11b/CD18 integrins. *PNAS*, 102:431–436, 2005.

N. C. Lau, L. P. Lim, E. G. Weinstein, and D. P. Bartel. An abundant class of tiny RNAs with probable regulatory roles in Caenorhabditis elegans. *Science*, 294(5543):858–862, Oct 2001.

D. D. Lee and H. S. Seung. Learning the parts of objects by non-negative matrix factorization. *Nature*, 401(6755):788–791, Oct 1999.

D. D. Lee and H. S. Seung. Algorithms for non-negative matrix factorization. *Advances in neural information processing systems*, pages 556–562, 2001.

R. C. Lee, R. L. Feinbaum, and V. Ambros. The C. elegans heterochronic gene lin-4 encodes small RNAs with antisense complementarity to lin-14. *Cell*, 75(5):843–854, Dec 1993.

S.-I. Lee and S. Batzoglou. Application of independent component analysis to microarrays. *Genome Biology*, 4(11):R76, 2003.

Y. Lee, C. Ahn, J. Han, H. Choi, J. Kim, J. Yim, J. Lee, P. Provost, O. Rådmark, S. Kim, and V. N. Kim. The nuclear RNase III Drosha initiates microRNA processing. *Nature*, 425(6956):415–419, Sep 2003.

B. Lehner and C. M. Sanderson. A protein interaction framework for human mRNA degradation. *Genome Res*, 14(7):1315–1323, Jul 2004.

F. Lejeune, X. Li, and L. E. Maquat. Nonsense-mediated mRNA decay in mammalian cells involves decapping, deadenylating, and exonucleolytic activities. *Mol Cell*, 12(3):675–687, Sep 2003.

E. Levine, B. Jacob, et al. Target-Specific and Global Effectors in Gene Regulation by MicroRNA. *Biophysical Journal*, 93(11):L52, 2007.

B. Lewis, I. Shih, M. Jones-Rhoades, D. Bartel, and C. Burge. Prediction of Mammalian MicroRNA Targets. *Cell*, 115(7):787–798, 2003.

S. Li, X. Hou, H. Zhang, and Q. Cheng. Learning spatially localized, parts-based representation. In *IEEE Computer Society Conference on Computer Vision and Pattern Recognition*, volume 1. IEEE Computer Society; 1999, 2001.

W. Liebermeister. Linear modes of gene expression determined by independent component analysis. *Bioinformatics*, 18:51–60, 2002.

L. P. Lim, M. E. Glasner, S. Yekta, C. B. Burge, and D. P. Bartel. Vertebrate microRNA genes. *Science*, 299(5612):1540, Mar 2003.

G. Liu, A. E. Loraine, R. Shigeta, M. Cline, J. Cheng, V. Valmeekam, S. Sun, D. Kulp, and M. A. Siani-Rose. Netaffx: Affymetrix probesets and annotations. *Nucleic Acids Res*, 31(1):82–86, Jan 2003.

K. Lock, J. Zhang, J. Lu, S. Lee, and P. Crocker. Expression of cd33-related siglecs on human mononuclear phagocytes, monocyte-derived dendritic cells and plasmacytoid dendritic cells. *Immunobiology*, 209:199–207, 2004.

D. J. Loegring, J. R. Drake, J. A. Banas, T. L. McNeal, D. G. M. Arthur, L. M. Webster, and M. R. Lennartz. Francisella tularensis LVS grown in macrophages has reduced ability to stimulate the secretion of inflammatory cytokines by macrophages in vitro. *Microb Pathog*, 41:218–225, 2006.

M. C. López, N. S. Duckett, S. D. Baron, and D. W. Metzger. Early activation of NK cells after lung infection with the intracellular bacterium, Francisella tularensis LVS. *Cell Immunol*, 232(1-2):75–85, 2004.

D. Lutter, K. Stadlthanner, F. Theis, E. W. Lang, A. Tomé, B Becker, and T. Vogt. Analyzing gene expression profiles with ICA. In *Proceedings of the 24th IASTED international conference on Biomedical engineering*, pages 25–30. ACTA Press Anaheim, CA, USA, 2006.

D. Lutter, P. Ugocsai, M. Grandl, E. Orso, F. Theis, E. W. Lang, and G. Schmitz. Analyzing M-CSF dependent monocyte/macrophage differentiation: expression modes and meta-modes derived from an independent component analysis. *BMC Bioinformatics*, 9:100, 2008.

D. Lutter, T. Langmann, P. Ugocsai, C. Moehle, E. Seibold, W. Splettstoesser, P. Gruber, E. W. Lang, and G. Schmitz. Analyzing time-dependent microarray data using independent component analysis derived expression modes from human macrophages infected with F. tularensis holartica. *Journal of Biomedical Informatics*, 2009. doi: 10.1016/j.jbi.2009.01.002.

S. Martin and R. Parton. Lipid droplets: a unified view of a dynamic organelle. *Nat Rev Mol Cell Biol.*, 7:373–378, 2006.

F. Martinez, S. Gordon, M. Locati, and A. Mantovani. Transcriptional profiling of the human monocyte-to-macrophage differentiation and polarization: new molecules and patterns of gene expression. *Journal of Immunology*, 177:7303–7311, 2006.

K. Y. Y. J. Masutani H, Bai J. Thioredoxin as a neurotrophic cofactor and an important regulator of neuroprotection. *Mol Neurobiol.*, 29:229–249, 2004.

T. Mathworks. MATLAB R2008a, Natick, Massachusetts: The MathWorks Inc, 2008. URL http://www.mathworks.com/.

J. S. Mattick and I. V. Makunin. Non-coding RNA. *Hum Mol Genet*, 15 Spec No 1:R17–R29, Apr 2006.

P. Mina-Osorio and E. Ortega. Signal regulators in FcR-mediated activation of leukocytes? *Trends Immunol.*, 25:529–35, 2004.

M. Minami, N. Kume, T. Shimaoka, and T. Kita. Expression of SR-PSOX, a Novel Cell-Surface Scavenger Receptor for Phosphatidylserine and Oxidized LDL in Human Atherosclerotic Lesions. *Arteriosclerosis, Thrombosis, and Vascular Biology*, 21:1796–1800, 2001.

D. Moazed. Small RNAs in transcriptional gene silencing and genome defence. *Nature*, 457 (7228):413–420, Jan 2009.

K. V. Morris, S. Santoso, A.-M. Turner, C. Pastori, and P. G. Hawkins. Bidirectional transcription directs both transcriptional gene activation and suppression in human cells. *PLoS Genet*, 4(11):e1000258, Nov 2008.

N. U. Nair and P. G. Sankaran. Characterization of the pearson family of distributions. *IEEE Transactions on Reliability*, 40:75–77, 1991.

S. J. Nelson, M. Schopen, A. G. Savage, J.-L. Schulman, and N. Arluk. The MeSH translation maintenance system: structure, interface design, and implementation. *Stud Health Technol Inform*, 107(Pt 1):67–69, 2004.

S. F. Newbury. Control of mRNA stability in eukaryotes. *Biochem Soc Trans*, 34(Pt 1):30–34, Feb 2006.

J. Nordberg and E. S. Arner. Reactive oxygen species, antioxidants, and the mammalian thioredoxin system. *Free Radic Biol Med.*, 31:1287–1312, 2001.

N. Osato, Y. Suzuki, K. Ikeo, and T. Gojobori. Transcriptional interferences in cis natural antisense transcripts of humans and mice. *Genetics*, 176(2):1299–1306, Jun 2007.

B. Østerud and E. Bjørklid. Role of monocytes in atherogenesis. *Physiol. Rev.*, 83:1069–1112, 2003.

E. Ostrakhovitch, P. Olsson, S. Jiang, and M. Cherian. Interaction of metallothionein with tumor suppressor p53 protein. *FEBS Lett.*, 580:1235–1238, 2006.

D. Park, A. Thomsen, C. Frevert, U. Pham, S. Skerrett, P. Kiener, and W. Liles. Fas (CD95) induces proinflammatory cytokine responses by human monocytes and monocyte-derived macrophages. *J. Immunol.*, 170:6209–6216, 2003.

R. Parker and U. Sheth. P bodies and the control of mRNA translation and degradation. *Mol Cell*, 25(5):635–646, Mar 2007.

R. Parker and H. Song. The enzymes and control of eukaryotic mRNA turnover. *Nat Struct Mol Biol*, 11(2):121–127, Feb 2004.

L. Peiser, S. Mukhopadhyay, and S. Gordon. Scavenger receptors in innate immunity. *Curr Opin Immunol.*, 14:123–128, 2002.

M. Pheasant and J. Mattick. Raising the estimate of functional human sequences. *Genome Research*, 17(9):1245, 2007.

F. J. Pixley and E. R. Stanley. CSF-1 regulation of the wandering macrophage: complexity in action. *Trends in Cell Biology*, 14:628–638, 2004.

K. Pruitt, T. Tatusova, and D. Maglott. NCBI reference sequences (RefSeq): a curated nonredundant sequence database of genomes, transcripts and proteins. *Nucleic Acids Research*, 35:D61, 2007.

J. Quackenbush. Computational Analysis of Microarray Data. *Nature*, 2:418 – 427, 2001.

J. Quackenbush. Computational approaches to analysis of DNA microarray data. *Yearb Med Inform*, pages 91–103, 2006.

B. J. Reinhart, E. G. Weinstein, M. W. Rhoades, B. Bartel, and D. P. Bartel. MicroRNAs in plants. *Genes Dev*, 16(13):1616–1626, Jul 2002.

M. W. Rhoades, B. J. Reinhart, L. P. Lim, C. B. Burge, B. Bartel, and D. P. Bartel. Prediction of plant microRNA targets. *Cell*, 110(4):513–520, Aug 2002.

A. Rodriguez, S. Griffiths-Jones, J. L. Ashurst, and A. Bradley. Identification of mammalian microRNA host genes and transcription units. *Genome Res*, 14(10A):1902–1910, Oct 2004.

F. P. Ross and S. L. Teitelbaum. $\alpha_v\beta_3$ and macrophage colony-stimulating factor: partners in osteoclast biology . *Immunological Reviews*, 208:88–105, 2005.

J. Ruby, C. Jan, and D. Bartel. Intronic microRNA precursors that bypass Drosha processing. *Nature*, 448(7149):83, 2007.

B. Ryan, N. O'Donovan, B. Browne, C. O'Shea, J. Crown, A. D. K. Hill, E. McDermott, N. O'Higgins, and M. J. Duffy. Expression of survivin and its splice variants survivin-2B and survivin-DeltaEx3 in breast cancer. *Br J Cancer*, 92(1):120–124, Jan 2005.

H. Saini, A. Enright, and S. Griffiths-Jones. Annotation of Mammalian Primary microRNAs. *BMC Genomics*, 9:564, 2008.

D. Sarkar, R. Parkin, S. Wyman, A. Bendoraite, C. Sather, J. Delrow, A. K. Godwin, C. Drescher, W. Huber, R. Gentleman, and M. Tewari. Quality assessment and data analysis for microRNA expression arrays. *Nucleic Acids Res*, 37(2):e17, Feb 2009.

R. Schachtner, D. Lutter, K. Stadlthanner, E. W. Lang, G. Schmitz, A. M. Tomé, and P. G. Vilda. Routes to identify marker genes for microarray classification. *Conf Proc IEEE Eng Med Biol Soc*, 2007:4617–4620, 2007a.

R. Schachtner, D. Lutter, F. J. T. Theis, E. W. Lang, G. Schmitz, A. M. Tomé, and P. G. Vilda. How to extract marker genes from microarray data sets. *Conf Proc IEEE Eng Med Biol Soc*, 2007:4215–4218, 2007b.

R. Schachtner, D. Lutter, P. Knollmüller, A. M. Tomé, F. J. Theis, G. Schmitz, M. Stetter, P. G. Vilda, and E. W. Lang. Knowledge-based gene expression classification via matrix factorization. *Bioinformatics*, 24(15):1688–1697, Aug 2008.

A. Schlicker, F. Domingues, J. Rahmenfuhrer, and T. Lengauer. A new measure for functional similarity of gene products based on Gene Ontology. *BMC Bioinformatics*, 7(1):302, 2006.

G. Schmitz and C. Buechler. ABCA1: regulation, trafficking and association with heteromeric proteins. *Ann Med.*, 34:334–47, 2002.

G. Schmitz and M. Grandl. Role of redox regulation and lipid rafts in macrophages during Ox-LDL-mediated foam cell formation. *Antioxid Redox Signal*, 9(9):1499–1518, Sep 2007.

B. Scholkopf and A. Smola. *Learning with kernels*. MIT press Cambridge, Mass, 2002.

L. Scorrano, S. A. Oakes, J. T. Opferman, E. H. Cheng, M. D. Sorcinelli, T. Pozzan, and S. J. Korsmeyer1dagger. BAX and BAK Regulation of Endoplasmic Reticulum Ca2+: A Control Point for Apoptosis. *Science*, 300:135 – 139, 2003.

M. Selbach, N. Thierfelder, Z. Fang, R. Khanin, and N. Rajewsky. Widespread changes in protein synthesis induced by microRNAs. *Nature*, 2008.

R. Shalgi, D. Lieber, M. Oren, and Y. Pilpel. Global and local architecture of the mammalian microRNA-transcription factor regulatory network. *PLoS Comput Biol*, 3(7):e131, Jul 2007.

C. Shi and D. Simon. Integrin signals, transcription factors, and monocyte differentiation. *Trends Cardiovasc Med.*, 16:146–152, 2006.

R. Singal and G. D. Ginder. DNA methylation. *Blood*, 93(12):4059–4070, Jun 1999.

A. Sobota, A. Strzelecka-Kiliszek, E. Gladkowska, K. Yoshida, K. Mrozinska, and K. Kwiatkowska. Binding of IgG-Opsonized Particles to FcγR Is an Active Stage of Phagocytosis That Involves Receptor Clustering and Phosphorylation. *Journal of Immunology*, 175: 4450–4457, 2005.

K. Stadlthanner, F. Theis, E. W. Lang, A. M. Tomé, W. Gronwald, and H. R. Kalbitzer. A matrix pencil approach to the blind source separation of artifacts in 2D NMR spectra. *Neural Information Processing - Letters and Reviews*, 1:103 – 110, 2003a.

K. Stadlthanner, A. Tomé, F. Theis, W. Gronwald, H. Kalbitzer, and E. Lang. Blind source separation of water artefacts in NMR spectra using a matrix pencil. *Proc. ICA 2003*, pages 167–172, 2003b.

K. Stadlthanner, F. Theis, E. W. Lang, A. M. Tomé, W. Gronwald, and H. R. Kalbitzer. Separation of water artefacts in 2D NOESY protein spectra using congruent matrix pencils. *Neurocomputing*, 2005.

K. Stadlthanner, D. Lutter, F. Theis, E. Lang, A. Tome, P. Georgieva, and C. Puntonet. Sparse Nonnegative Matrix Factorization with Genetic Algorithms for Microarray Analysis. In *Neural Networks, 2007. IJCNN 2007. International Joint Conference on*, pages 294–299, 2007.

G. Stefani and F. Slack. Small non-coding RNAs in animal development. *Nature Reviews Molecular Cell Biology*, 9(3):219, 2008.

D. Stekel. *Microarray Bioinformatics*. Cambridge University Press, 2003.

L. Stephens, C. Ellson, and P. Hawkins. Roles of PI3Ks in leukocyte chemotaxis and phagocytosis. *Curr. Opin. Cell Biol*, 14:203–213, 2002.

G. Storz. An expanding universe of noncoding RNAs. *Science*, 296(5571):1260–1263, May 2002.

A. Sturn, J. Quackenbush, and Z. Trajanoski. Genesis: Cluster analysis of microarray data. *Bioinformatics*, 18(1):207–208, Jan 2002.

J. A. Swanson and A. D. Hoppe. The coordination of signaling during Fc receptor-mediated phagocytosis. *Journal of Leukocyte Biology*, 76:1093–1103, 2004.

F. J. Theis. *Mathematics in independent component analysis*. PhD thesis, Universität Regensburg, 2002.

F. J. Theis and E. W. Lang. Formalization of the two-step approach to overcomplete BSS. *Proc. of SIP 2002*, pages 207–212, 2002.

F. J. Theis, E. W. Lang, and C. G. Puntonet. A geometric algorithm for overcomplete linear ICA. *Neurocomputing*, 56:381–398, 2004a.

F. J. Theis, A. Meyer-Baese, and E. W. Lang. Second-order blind source separation based on multi-dimensional autocovariances. In A. P. C. G. Puntonet, editor, *Lecture Notes in Computer Science (LNCS 3195), Proc. ICA'2004*, pages 726 – 733, Berlin, 2004b. Springer Verlag.

F. J. Theis, P. Gruber, I. R. Keck, and E. W. Lang. Functional MRI analysis by a novel spatiotemporal ICA algorithm. *Lecture notes in computer science*, 3696:677, 2005.

A. M. Tomé, A. R. Teixeira, E. W. Lang, K. Stadlthanner, and A Rocha. Blind Source Separation Using Time-delayed Signals. In *Proceedings of the International Joint Conference on Neural Networks, IJCNN'2004*, volume CD, Budapest, Hungary, 2004.

M. Toshiyuki and J. C. Reed. Tumor suppressor p53 is a direct transcriptional activator of the human bax gene. *Cell*, 80:293–299, 1995.

J. Tsang, J. Zhu, and A van Oudenaarden. MicroRNA-mediated feedback and feedforward loops are recurrent network motifs in mammals. *Mol Cell*, 26(5):753–767, Jun 2007.

V. G. Tusher, R. Tibshirani, and G. Chu. Significance analysis of microarrays applied to the ionizing radiation response. *Proc Natl Acad Sci U S A*, 98(9):5116–5121, Apr 2001.

S. Vasudevan, Y. Tong, and J. A Steitz. Switching from repression to activation: microRNAs can up-regulate translation. *Science*, 318(5858):1931–1934, Dec 2007.

O. Vieira, R. Botelho, and S. Grinstein. Phagosome maturation aging gracefully. *Biochem. J.*, 366:689–704, 2002.

R. Vigario, V. Jousmäki M. Hämäläinen, R. Hari, and E. Oja. Independent component analysis for identification of artifacts in Magnetoencephalographic recordings. In *Advances in Neural Information Processing Systems 10 (NIPS'1997)*, pages 229 – 235, Cambridge, Masachusetts, 1997. MIT Press.

T. Wada and J. M. Penninger. Mitogen-activated protein kinases in apoptosis regulation. *Oncogene*, 23:2838–2849, 2004.

Y. Wang, M. Qiao, J. Mieyal, L. Asmis, and R. Asmis. Molecular mechanism of glutathione-mediated protection from oxidized low-density lipoprotein-induced cell injury in human macrophages: role of glutathione reductase and glutaredoxin. *Free Radic Biol Med.*, 41:775–85, 2006a.

Y. Wang, M. M. Zeigler, G. K. Lam, M. G. Hunter, T. D. Eubank, V. V. Khramtsov, S. Tridandapani, C. K. Sen, and C. E. Marsh. The Role of the NADPH Oxidase Complex, p38 MAPK, and Akt in Regulating Human Monocyte/Macrophage Survival. *American Journal of Respiratory and Critical Care Medicine*, 36:68–77, 2006b.

Z. Wang, M. Gerstein, and M. Snyder. RNA-Seq: a revolutionary tool for transcriptomics. *Nat Rev Genet*, 10(1):57–63, Jan 2009.

B. Wightman, I. Ha, and G. Ruvkun. Posttranscriptional regulation of the heterochronic gene lin-14 by lin-4 mediates temporal pattern formation in C. elegans. *Cell*, 75(5):855–862, Dec 1993.

C. Wilson and C. Miller. Simpleaffy: a BioConductor package for Affymetrix Quality Control and data analysis, 2005.

D. Wilson, V. Charoensawan, S. K. Kummerfeld, and S. A. Teichmann. DBD–taxonomically broad transcription factor predictions: new content and functionality. *Nucleic Acids Res*, 36 (Database issue):D88–D92, Jan 2008.

K. Yang and J. C. Rajapakse. ICA gives higher-order functional connectivity of brain. *Neural Information Processing - Letters and Reviews*, 2:27 – 32, 2004.

K. Y. Yeung and W. L. Ruzzo. Principal component analysis for clustering gene expression data. *Bioinformatics*, 17(9):763–774, Sep 2001.

S. Ying and S. Lin. Intronic microRNAs. *Biochemical and Biophysical Research Communications*, 326(3):515–520, 2005.

X. Yu, J. Lin, D. J. Zack, J. T. Mendell, and J. Qian. Analysis of regulatory network topology reveals functionally distinct classes of microRNAs. *Nucleic Acids Res*, 36(20):6494–6503, Nov 2008.

P. D. Zamore, T. Tuschl, P. A. Sharp, and D. P. Bartel. RNAi: double-stranded RNA directs the ATP-dependent cleavage of mRNA at 21 to 23 nucleotide intervals. *Cell*, 101(1):25–33, Mar 2000.

Y. Zeng and B. R. Cullen. Sequence requirements for microRNA processing and function in human cells. *RNA*, 9(1):112–123, Jan 2003.

Y. Zhu, T. Kalbfleisch, M. D. Brennan, and Y. Li. A MicroRNA gene is hosted in an intron of a schizophrenia-susceptibility gene. *Schizophr Res*, 109(1-3):86–89, Apr 2009.

Die VDM Verlagsservicegesellschaft sucht für wissenschaftliche Verlage abgeschlossene und herausragende

Dissertationen, Habilitationen, Diplomarbeiten, Master Theses, Magisterarbeiten usw.

für die kostenlose Publikation als Fachbuch.

Sie verfügen über eine Arbeit, die hohen inhaltlichen und formalen Ansprüchen genügt, und haben Interesse an einer honorarvergüteten Publikation?

Dann senden Sie bitte erste Informationen über sich und Ihre Arbeit per Email an *info@vdm-vsg.de*.

Sie erhalten kurzfristig unser Feedback!

VDM Verlagsservicegesellschaft mbH
Dudweiler Landstr. 99　　　　　　Telefon +49 681 3720 174
D - 66123 Saarbrücken　　　　　　Fax　　　+49 681 3720 1749

www.vdm-vsg.de

Die VDM Verlagsservicegesellschaft mbH vertritt

Printed by Books on Demand GmbH, Norderstedt / Germany